Modeling and Simulation Based Analysis in Reliability Engineering

Advanced Research in Reliability and System Assurance Engineering

Series Editor:
Mangey Ram
*Professor, Department of Mathematics, Computer Science & Engineering,
Graphic Era Deemed to be University, Dehradun, India*

Modeling and Simulation Based Analysis
in Reliability Engineering
Mangey Ram

Reliability Engineering: Theory and Applications
Ilia Vonta and Mangey Ram

Reliability Engineering: Methods and Applications
Mangey Ram

System Reliability Management: Solutions and Technologies
Adarsh Anand and Mangey Ram

Modeling and Simulation Based Analysis in Reliability Engineering

Edited by
Mangey Ram

CRC Press
Taylor & Francis Group
Boca Raton London New York

CRC Press is an imprint of the
Taylor & Francis Group, an **informa** business

MATLAB® is a trademark of The MathWorks, Inc. and is used with permission. The MathWorks does not warrant the accuracy of the text or exercises in this book. This book's use or discussion of MATLAB® software or related products does not constitute endorsement or sponsorship by The MathWorks of a particular pedagogical approach or particular use of the MATLAB® software.

CRC Press
Taylor & Francis Group
6000 Broken Sound Parkway NW, Suite 300
Boca Raton, FL 33487-2742

First issued in paperback 2020

© 2019 by Taylor & Francis Group, LLC
CRC Press is an imprint of Taylor & Francis Group, an Informa business

No claim to original U.S. Government works

ISBN-13: 978-1-138-57021-4 (hbk)
ISBN-13: 978-0-367-78105-7 (pbk)

This book contains information obtained from authentic and highly regarded sources. Reasonable efforts have been made to publish reliable data and information, but the author and publisher cannot assume responsibility for the validity of all materials or the consequences of their use. The authors and publishers have attempted to trace the copyright holders of all material reproduced in this publication and apologize to copyright holders if permission to publish in this form has not been obtained. If any copyright material has not been acknowledged please write and let us know so we may rectify in any future reprint.

Except as permitted under U.S. Copyright Law, no part of this book may be reprinted, reproduced, transmitted, or utilized in any form by any electronic, mechanical, or other means, now known or hereafter invented, including photocopying, microfilming, and recording, or in any information storage or retrieval system, without written permission from the publishers.

For permission to photocopy or use material electronically from this work, please access www.copyright.com (http://www.copyright.com/) or contact the Copyright Clearance Center, Inc. (CCC), 222 Rosewood Drive, Danvers, MA 01923, 978-750-8400. CCC is a not-for-profit organization that provides licenses and registration for a variety of users. For organizations that have been granted a photocopy license by the CCC, a separate system of payment has been arranged.

Trademark Notice: Product or corporate names may be trademarks or registered trademarks, and are used only for identification and explanation without intent to infringe.

Library of Congress Cataloging-in-Publication Data

Names: Ram, Mangey, editor.
Title: Modeling and simulation based analysis in reliability engineering/ editor, Mangey Ram.
Description: Boca Raton, FL : CRC Press/Taylor & Francis Group, 2018. |
Series: Advanced research in reliability and system assurance engineering |
"A CRC title, part of the Taylor & Francis imprint, a member of the
Taylor & Francis Group, the academic division of T&F Informa plc."
Includes bibliographical references and index.
Identifiers: LCCN 2018011971| ISBN 9781138570214 (hardback : acid-free paper) |
ISBN 9780203703717 (ebook)
Subjects: LCSH: Reliability (Engineering)--Data processing.
Classification: LCC TA169 .M637 2018 | DDC 620/.00452011--dc23
LC record available at https://lccn.loc.gov/2018011971

Visit the Taylor & Francis Web site at
http://www.taylorandfrancis.com

and the CRC Press Web site at
http://www.crcpress.com

Contents

Preface vii
Acknowledgments ix
Author xi
List of Contributors xiii

1. Modeling the Assessment and Monitoring of Reliability of the
   Condensation Thermal Power Plants (Co-TPP) 1
   Zdravko Milovanović, Svetlana Dumonjić-Milovanović, and Ljubiša Papić

2. Non-exponential Distributions in Reliability Modeling of PMS:
   Approximation and Simulation Approaches 47
   Xiang-Yu Li, Jun Hu, Hong-Zhong Huang, and Yan-Feng Li

3. Optimal Periodic Software Rejuvenation Policies in Discrete
   Time—Survey and Applications 81
   Tadashi Dohi, Junjun Zheng, and Hiroyuki Okamura

4. Potential Applications of Multivariate Analysis for Modeling
   the Reliability of Repairable Systems—Examples Tested 109
   Miguel Angel Navas, Carlos Sancho, and Jose Carpio

5. Phased Mission Systems—Modeling and Reliability 141
   Kanchan Jain, Isha Dewan, and Monika Rani

6. Bayesian Inference on General-Order Statistic Models 163
   Aniket Jain, Biswabrata Pradhan, and Debasis Kundu

7. Large-Scale Reliability-Redundancy Allocation Optimization
   Problem Using Three Soft Computing Methods 183
   Mohamed Arezki Mellal and Edward J. Williams

8. A New Distribution-Free Reliability Monitoring Scheme:
   Advances and Applications in Engineering 199
   Ioannis S. Triantafyllou

9. Modeling and Simulation of a Sustainable Hybrid Energy
   System under Changing Power Reliability Index at User End 215
   Anurag Chauhan

10. **Signature Reliability of *k*-out-of-*n* Sliding Window System** 233
    Akshay Kumar and Mangey Ram

11. **Modeling Reliability of Component-Based Software Systems** 249
    Preeti Malik, Lata Nautiyal, and Mangey Ram

12. **Reliability and Fault Tolerance Modeling of Multiphase Traction Electric Motors** 267
    Ilia Frenkel, Lev Khvatskin, Ehud Ikar, Igor Bolvashenkov, Hans-Georg Herzog, and Anatoly Lisnianski

Index 297

Preface

Recently, developments in reliability engineering have become the most challenging and demanding area of research. System reliability engineering has become a greater apprehension in modern years because of high-tech industrial processes with most engineering systems today having ever-increasing levels of complexity. Reliability theory forms the common foundation of engineering as it evolves and develops. The book *"Modeling and Simulation Based Analysis in Reliability Engineering"* engrossed on a comprehensive range of modeling and simulation in reliability engineering. Topics of focus include:

- Modeling the assessment and monitoring of reliability of the condensation thermal power plants
- Non-exponential distributions in reliability modeling of PMS: approximation and simulation approaches
- Optimal periodic software rejuvenation policies in discrete time
- Potential applications of multivariate analysis for modeling reliability of repairable systems
- Phased mission systems
- Bayesian inference on general-order statistic models
- Large scale reliability-redundancy allocation optimization problem
- The new distribution-free reliability monitoring scheme: advances and applications in engineering
- Modeling and simulation of a sustainable hybrid energy system under changing power reliability index
- Signature reliability of k-out-of-n sliding window system
- Modeling reliability of component-based software systems
- Reliability and fault tolerance modeling of multiphase traction electric motors

The book is meant for those who want to take reliability engineering as a subject of study. This book is very useful to the undergraduate and postgraduate students of engineering; and engineers, research scientists, and academicians involved in the engineering sciences.

MATLAB® is a registered trademark of The MathWorks, Inc. For product information, please contact:

The MathWorks, Inc.
3 Apple Hill Drive
Natick, MA 01760-2098 USA
Tel: 508-647-7000
Fax: 508-647-7001
E-mail: info@mathworks.com
Web: www.mathworks.com

Acknowledgments

The editor acknowledges CRC Press for this opportunity and professional support. My special thanks to Ms. Cindy Renee Carelli, executive editor, CRC Press—Taylor & Francis Group for the excellent support she provided to me in completing this book. Thanks to Ms. Renee Nakash, editorial assistant to Ms. Carelli for her follow-up and aid. Also, I would like to thank all the chapter authors and reviewers for their availability for this work.

Mangey Ram
Graphic Era Deemed to be University, India

Author

Mangey Ram received his PhD in Mathematics and a minor in Computer Science from G. B. Pant University of Agriculture and Technology, Pantnagar, India, in 2008. He has been a faculty member for approximately 10 years and has taught several core courses in pure and applied mathematics at undergraduate, postgraduate, and doctorate levels. He is currently a professor at Graphic Era Deemed to be University, Dehradun, India. Before joining Graphic Era, he was the deputy manager (probationary officer) with Syndicate Bank for a short period. He is editor-in-chief of *International Journal of Mathematical, Engineering and Management Sciences* and the guest editor & member of the editorial board of various journals. He is also regular reviewer for international journals, including IEEE, Elsevier, Springer, Emerald, John Wiley, Taylor & Francis Group, and many other publishers. He has published around 130 research publications in IEEE, Taylor & Francis Group, Springer, Elsevier, Emerald, World Scientific, and many other national and international journals of repute and also presented his works at national and international conferences. His fields of research are reliability theory and applied mathematics. Dr. Ram is a senior member of the IEEE, life member of Operational Research Society of India, Society for Reliability Engineering, Quality and Operations Management in India, Indian Society of Industrial and Applied Mathematics, member of International Association of Engineers in Hong Kong, and Emerald Literati Network in the UK. He has been a member of the organizing committee of a number of international and national conferences, seminars, and workshops. He was conferred with "Young Scientist Award" by the Uttarakhand State Council for Science and Technology, Dehradun, in 2009. He was awarded the "Best Faculty Award" in 2011, and recently, the Research Excellence Award in 2015 for his significant contribution in academics and research at Graphic Era.

List of Contributors

Igor Bolvashenkov
Institute of Energy Conversion Technology
Technical University of Munich
Munich, Germany

Jose Carpio
Spanish National Distance Education University
Madrid, Spain

Anurag Chauhan
Department of Electrical Engineering
Rajkiya Engineering College
Banda, India

Isha Dewan
Statistics and Mathematics Unit
Indian Statistical Institute
Delhi, India

T. Dohi
Hiroshima University
Higashi-Hiroshima, Japan

Svetlana Dumonjić-Milovanović
Partner Engineering Ltd.
Banja Luka, Republic of Srpska

Ilia Frenkel
Center for Reliability and Risk Management, Industrial Engineering and Management Department
SCE - Shamoon College of Engineering
Beer sheva, Israel

Hans-Georg Herzog
Institute of Energy Conversion Technology
Technical University of Munich
Munich, Germany

Jun Hu
School of Foreign Languages
University of Electronic Science and Technology of China
Chengdu, Sichuan, 611731, P. R. China

Hong-Zhong Huang
Center for System Reliability and Safety
University of Electronic Science and Technology of China
Chengdu, Sichuan, 611731, P. R. China

Ehud Ikar
Center for Reliability and Risk Management, Industrial Engineering and Management Department
SCE - Shamoon College of Engineering
Beer sheva, Israel

Aniket Jain
Acellere Software Pvt. Ltd.
Pune, India

Kanchan Jain
Department of Statistics
Panjab University
Chandigarh, India

xiii

Lev Khvatskin
Center for Reliability and Risk Management, Industrial Engineering and Management Department
SCE - Shamoon College of Engineering
Beer sheva, Israel

Akshay Kumar
Tula's Institute, The Engineering and Management College
Dehradun, India

Debasis Kundu
Department of Mathematics and Statistics
Indian Institute of Technology
Kanpur, India

Xiang-Yu Li
Center for System Reliability and Safety
University of Electronic Science and Technology of China
Chengdu, Sichuan, 611731, P. R. China

Yan-Feng Li
Center for System Reliability and Safety
University of Electronic Science and Technology of China
Chengdu, Sichuan, 611731, P. R. China

Anatoly Lisnianski
The Israel Electric Corporation
Haifa, Israel

Preeti Malik
Department of Computer Application
Graphic Era Deemed to be University
Dehradun, India

Mohamed Arezki Mellal
M'Hamed Bougara University
Boumerdes, Algeria

Miguel Angel Navas
Spanish National Distance Education University
Madrid, Spain

Zdravko Milovanović
Faculty of Mechanical Engineering
University of Banja Luka
Banja Luka, Republic of Srpska

Lata Nautiyal
Department of Computer Application
Graphic Era Deemed to be University
Dehradun, India

H. Okamura
Hiroshima University
Higashi-Hiroshima, Japan

Ljubiša Papić
The Research Center of Dependability and Quality Management DQM
Prijevor, Serbia

Biswabrata Pradhan
Statistical Quality Control and Operation Research Unit
Indian Statistical Institute
Kolkata, India

List of Contributors

Mangey Ram
Department of Mathematics
Computer Science and Engineering
Graphic Era Deemed to be
  University
Dehradun, India

Monika Rani
TBRL, Defence Research
  Development Organisation
Chandigarh, India

Carlos Sancho
Spanish National Distance
  Education University
Madrid, Spain

Ioannis S. Triantafyllou
Department of Computer
  Science & Biomedical
  Informatics
University of Thessaly
Lamia, Greece

Edward J. Williams
University of Michigan
Dearborn, Michigan

J. Zheng
Hiroshima University
Higashi-Hiroshima, Japan

1

Modeling the Assessment and Monitoring of Reliability of the Condensation Thermal Power Plants (Co-TPP)

Zdravko Milovanović and Svetlana Dumonjić-Milovanović
University of Banja Luka

Ljubiša Papić
University of Kragujevac

CONTENTS

1.1 Introduction 1
1.2 Models for Predicting the Reliability of Complex Technical Systems 3
1.3 Mathematical Models of the Growth of Reliability of System 12
1.4 Reliability Assessment Methods 16
1.5 Indicators of Reliability of Co-TPP 18
1.6 Failures and Damages during the Operation of Co-TPP 26
1.7 Modified Method for Assessment of the Optimal Reliability of Co-TPP 31
1.8 System of Maintenance with Entries and Exits Toward Environment 35
1.9 Overview of Maintenance Activities in TPP 38
1.10 Unsolved Tasks and Directions of Further Developing Models for Forecasting Reliability of CTSs 38
1.11 Conclusion 41
References 43

1.1 Introduction

The development and use of condensation thermal power plants (Co-TPP) is characterized today by great complexity, regardless of whether a technological scheme or built-in equipment is observed. On the other hand, large energy plants with new or improved solutions can be built only in case they have a high degree of safety and reliability and when they fully meet the applicable environmental criteria. Any disturbance in the operating mode of such a plant or reduction of its power also affects the electric power system (increase in the reserves of production capacities in it, uneven supply of electricity to consumers, etc.). Reliability assessment methods are mainly based on results of experiments on the set of system components based on observations of parameters of number and/or time of failure. In order to determine the reliability of components, it is necessary to either conduct some long-term and very expensive tests on a very large number of samples under special operating modes collect the data from exploitations, which is very risky. At the same time, the choice of general mathematical methods is especially important, due to the different shapes of the curves which quantitatively define the reliability with different functions of the failure rate and the great dependency of such curves on the change in the operating mode of the components and environmental conditions. In an attempt to overcome the above problems, we find that the introduction of approximate calculations gives an overview not only of the basic characteristics of the reliability of the observed system as a whole, but also insufficiently exact final parameters, due to a whole series of larger or smaller approximations, as well as the inability to take into account all the existing influences (development of new technologies, specificities of new disorders, etc.). On the other hand, the calculation of reliability of a complex system represents only the first initial phase of verification of quantitative features, that is, the very formed hypothesis in which we have more or less confidence. Their final acceptance or refusal represents the verification of reliability through the control of certain quantitative indicators of the system for the set technical conditions of operation. For these reasons, the alternative concepts, such as reliability control or hypothesis testing, have been often used in the literature to verify the reliability. Diagnostics, the evaluation of the state of elements of the facility together with tracking the progression of its aging, is very complex, responsible, and expensive task which demands educated personnel and modern diagnostic equipment. Diagnostic equipment available in the market is filled with diversity and methods used for diagnostic purposes are not generally accepted. The results of conducted diagnostic controls do not always give full answers, so that they are often limited to the monitoring of trend of change of observed diagnostic values. Consequently, experience becomes an unavoidable and immeasurable element of diagnostics. Experience itself is of course only possible to be acquired through the work and usage of diagnostic equipment, but it is also necessary to keep in mind the cost of experience acquirement in relation to a risk of investment in testing equipment. Development of diagnostic methods is intense regarding both field and laboratory methods but efforts to provide more and more cost-effective application methods are highly required. By developing new technologies and through the application of modern equipment and tools for monitoring of present state and diagnostics of the primary gear, the cost of routine maintenance that makes possible to recognize priorities of intervention maintenance can be decreased.

The methodology of maintenance regarding reliability also includes analysis of failure in the process of decision-making when maintenance is in question. A major problem in the early stage of development of maintenance strategies was analysis of reliability of complex energetic technical systems. Developing the aero-industry and introducing tracking of certain parameters during the operation (condition monitoring) was the basis of maintenance of technical systems (according to the condition).

1.2 Models for Predicting the Reliability of Complex Technical Systems

Today, the most frequently used models for predicting the reliability of technical systems are based on stochastic or statistical analysis, including Markov chains (processes), Poisson processes, Bayes method, state-based models, Monte Carlo simulations, and combinations of these models [1,2]. As their use in the analysis of complex energy systems is accompanied by significant limitations (most of these models focus only on mean time to failure (MTTF) or the expected number of failures for a given technical system), a smaller number of models regarding imperfect preventative maintenance actions have been developed (when the maintenance action fails to increase reliability level to 100%), as well as their integration with block diagrams [3]. One model for predicting the reliability of complex systems with a block diagram is the split system approach (SSA) model [4]. The researches related to the interactions between individual parts of technical systems and their failure or reliability prediction using a very low number of data about failures are also very limited and their application is very rare. A larger number of researches have been published about the technology of maintenance and models of reliability of technical systems. Predicative maintenance strategy, that is, condition based maintenance that belongs to the third generation of maintenance, carries out maintenance actions based on the condition of the parts or the entire system. As a result, the aims of this strategy are a higher level of reliability and availability of the plant, greater safety of operation, better product quality, longer lifetime of equipment, etc. The next strategy is a proactive maintenance strategy, with the aim not only of preventing the failure of the system, but also of setting the conditions to avoid or minimize the consequences of failure itself in the event of its occurrence. Additional optimization is the goal of maintenance, while prediction of reliability and risk assessment are the basis for making optimum maintenance decisions. The philosophy and concept of maintenance relate to the effective implementation of reliability assessment models, their implementation methodology, and strategic policies at the level of business management structures. Maintenance and reliability studies published so far can be classified into one of the categories shown in Figure 1.1, which provides an overview of the research in the domain of science on the maintenance and reliability of complex technical systems (CTSs). Below is an overview of several important conceptual approaches to reliability research as a CTS.

The power plant reliability index, which is decomposed and graphically presented in the research of Nikhil Dev et al., is a solid attempt to solve a crucial problem in such CTSs that are preventively maintained in the sense that it is possible to compensate for a small number of data on failures [5].

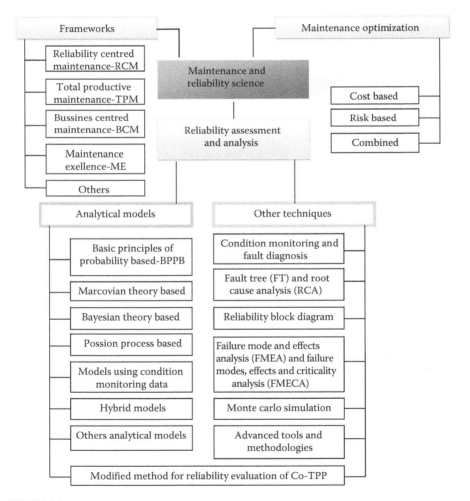

FIGURE 1.1
The overview of research in domain of maintenance and reliability science of technical systems. (Based on Sun [4] and Milovanović [11].)

Since the model relies heavily on the empirical experience and does not have enough objective data, its application is limited to more efficient energy systems. Exploitation systems and systems that can be decomposed into a number of components after a long period of exploitation by failure cannot be used to apply this reliability model. More importantly, interactions of failures of the system components are not predicted by the model. The result of the model is the real time reliability index (RTRI), which in fact represents the value of reliability at a time point of calculation.

A larger number of researches rely on Monte Carlo simulations, such as those of Naess et al., attempting to bypass the problems of mathematically complex or undetected rules of reliability modeling [6]. This method gives space for successful overcoming of such a task, but the models lack time dynamics and/or preventive maintenance actions. Thereby, the models have lost the possibilities of real implementation, especially when it is necessary to apply them to systems of a more complex structure.

The research of Weber and Jouffe combines several methods, such as the method of analysis of the fault tree, the Bayesian network, and the Markov chains [7]. This research is focused on modeling the reliability of production processes in CTSs. This also means higher requirements in the domain of reliability modeling, especially since this model has taken into account the existence of preventive maintenance actions as well as the time dynamics of reliability. The thing which is not complied with the real circumstances are the interactions of failures and the possibility of applying the model to CTSs decomposed into a large number of components due to the exponential increase in the number of combinations in the model, and the calculations for individual cases cannot be done within a reasonable time text.

Moazzami et al. gave assessments of the reliability of busbars in the power plant using Monte Carlo simulations; however, the time dynamics of movements of reliability assessment is missing, and no cases of repairs having a real impact on the system reliability are foreseen [8]. Also, reliability assessed based on parameters has been partially observed without an integral model.

In his dissertation, Sun developed several models relying on the existing ones such as SSA and analytical model for interactive failures (AMIF) [4]. The new developed model is the extended split system approach (ESSA) which involves the case of a small number of system failures, preventive maintenance actions, imperfect fixes, interactions of failure, and time dynamics. The model also predicts the existence of cascading failures as the possibility of applying different distributions of failures. The model is focused on making decisions on timely preventive but not corrective maintenance actions, so that the combined maintenance is excluded from the model. The model allows the reliability of each component of the system to be specifically considered, and preventive maintenance actions are defined according to each component of the system separately. The interactions of the components are defined by the matrix of interactive coefficients where the impact of the failure of one component on the failure of the other component is constant over time. After the preventive failures of components, the change of characteristics of theoretical distribution of failures of the system components where there would be no preventive maintenance actions is possible. However, the heuristic approach to determining the coefficients of interaction has remained a weak point in the model and made it more difficult for application. The ESSA model does not have clearly stated algorithms, although it is clear that they can be defined. The model also has predefined times of preventive maintenance actions and the implementation of maintenance actions in terms of reliability is not part of the model, although it is a reality in case of maintenance based on reliability [4].

The research done by Petrović et al. is based on the Markov process, whereby the optimal preventive maintenance time is established by maximization of the system availability. The model is stochastic and considers maintenance actions and their effects. The model is an applied transport system that was decomposed into two components. Such a model can easily be oriented to reliability; an answer about the times of optimal maintenance actions can be provided based on its criteria. However, in order to be applied to structurally complex technical systems with a large number of components, this model would have to undergo significant changes and many issues have to be solved. The model also neither anticipates an imperfect maintenance action nor an interaction of failures of system components [9].

Milošević, in one of the conclusions during the research within his dissertation, proposed an integral model for ensuring the reliability of CTSs, such as thermal power plants [10]. This model was based on the assumption that by decomposition every system component can be subject to the separate process of simulation, that is, a certain group of components, when necessary, are simulated together according to the reliability, which is the case with interaction of failures of components. The development of this model started from the collection, arrangement, classification of data, and decomposition of the system on the database. After the decomposition, the reliability of irreparable system components was modeled, then corrective maintenance actions were introduced, and then the preventive, so that the model included the reliability of the components that were maintained in combination. By following the assumption of perfect and imperfect maintenance actions, where after the maintenance actions the reliability follows, two models for predicting the reliability of components were developed. In case of the model of imperfectly maintained components, the problem of assessment of parameters of reliability for the theoretical distribution, to which failures of components due to successive maintenance actions are subject, appeared. Since this problem was complex, it was solved by Monte Carlo simulation on concrete examples, and a further simulation based on the parameters obtained was developed. The greatest challenge was the development of reliability models with the existence of interactive failures. It also included preventive maintenance actions, which are often more frequent than corrective, and it was necessary to develop methods for diagnosis and quantification of failure interaction. For this purpose, software solutions were used to analyze the data on system failure. Preventive maintenance actions during downtime were also included as a reality in the model itself. Regression analysis established the interdependency of some component failures with the interactions of others. It was done using a software package that identified the potential mutual interaction of failures of all components. Based on this, an interaction matrix was developed that was the basis for further modeling after calculating the extent of the change of reliability of the affected after a certain interfailure of affecting components. This finally resulted in development of the proposed model.

Within the framework of his dissertation, Milovanović emphasizes the importance of defining and forecasting the reliability indicators in case of preventive maintenance of complex systems [11]. Corrective actions in terms of further risks of failures and prevention of major damage are possible only if they are based on timely assessments. The optimum management of the thermal power plant system should be based on the assessment and complex optimization of the reliability indicators, depending on the way they are secured and the hierarchical level of details of the system as a whole, as well as the current stages of the life cycle of the plant. For these reasons, the optimization process includes basic structural, parametric, and constructive solutions related to the thermal power plant system through the change of its most important characteristics: energy efficiency, maneuverability, reliability, and economic efficiency as a whole. The complex of optimization goals ends with the overall selection of reliability indicators and possible manners for their provision, given the already established rules regarding the higher hierarchical level of the electric power system. The proposed modified method for assessment of optimal reliability of the system of condensation thermal power plant provides a good basis for further work on its development and improvement of the accuracy of the assessed values, with the introduction of technical diagnostics and the modern information and management system. The maintenance costs incurred are minimally possible for each specific situation. The dependency of the cost of electricity generation in the thermal power plant from the level of reliability should be considered from two aspects: thermal power plants and users. In both cases, the point of minimum costs determines the optimal reliability of both the thermal power plant and the user.

The Monte Carlo method, as a general method for solving problems in different fields of science, is based on the use of random numbers and probability theory [12]. It is used to simulate physical phenomena and solve complex problems. The final solution of a system of equations that describes the relations among particular phenomena is usually based on random sampling of relations and interactions, with a large number of repetitions or calculations. In this regard, the use of this method is one of the best examples of using computers as a research tool, in solving problems depending on their formulation in a statistical and random environment, that is, in situations where physical experiments are either impractical (risky) or too expensive. The very essence of application of the concept of the probability theory within the Monte Carlo method is to find solutions to physical problems often not related to probability or reliability. The direct simulation of the Monte Carlo method has several steps, which can be defined as follows: defining the basic settings of the problem faced by the analyst, the lack of important prerequisites for performance of necessary experiments (the technological process does not allow it, too expensive process, too risky job requiring the work on the boundaries of the criterion function, lack of time to perform the experiment, etc.), it is not possible to obtain an exact mathematical expression (mathematical model) which would adequately describe the process, and based on which a solution would be found within the limits of the allowed error; without performing an analytical solution, a random process defining the solution of the problem is defined, it is simulated on the computer, and the parameters for its solution are assessed, with the definition of allowed error and the required number of repetitions of the process itself (the distribution of parameters of different elements of a CTS selection of a random sample of each element, selection of their basic and supplementary parameters, combination of these samples and obtaining of the reliability of the technical system as a whole). For the calculation of the reliability of CTSs using the Monte Carlo method, the reliability or unreliability of each element is represented by a series of random numbers, while the choice of numbers is done in a sequence, so that each successive number is another success or a lack of success (failure). In addition, a computer, in which the database a program for generating random numbers is already installed, is used as a tool. By the appropriate combination of results obtained from the selection and interpretation for each element of the technical system, simulation of the technical system as a whole is done. The very program installed on the computer functions on the basis of logical diagram and description of operations on the system, as well as the existing functional connections of the system elements. The further course of the procedure is determined based on the previous starting condition (the condition of the system in operation and the condition of the system in failure). In the case when the number represents a success or the correct condition (operational condition), the group of random numbers for the next element in the same logical input is set at the input, and the new value of the random number is determined depending on whether the element is a success or a lack of success (failure). The process continues until the failure or breakdown occurs, which omits this path from the logic diagram and automatically returns the activity to the closest connection of the other parallel path. Thereby, the program uses the appropriate random number for the first element in the parallel path. If in this case the selection simulates success for the first element, then the random number set for the second element in a parallel path is used to determine the success or failure of that element. The process continues until a successful path is found, which indicates the success of the technical system as a whole or until the failure is simulated on all possible paths. The speed of the process depends on the number of elements of the technical system and the degree of its complexity, as well as on the reliability of the elements of the system itself. In cases where the preparation of the computer program shows certain simplicity, it is recommended to develop a mathematical model and apply certain analytical expressions and solve them. There are certain groups of factors influencing the applicability of the Monte Carlo method in order to determine the reliability of CTSs, and the three more important factors are emphasized. Redundancy of many technical systems is sequential, and in some cases it is "actively parallel," meaning that the failure time of one element affects the reliability of its sequential redundant replacement. A set of random numbers which represents the success or failure of a redundant element is not a constant, but is a function of a special random number representing the failure time of an element in the first path. Each number of the first group, which represents the relation of the first element, also determines the group to be used for the redundant element [an alternative is that the second group of random numbers remains the same, but that the interpretation of each value as a success (operational condition) or a lack of success (failure) varies as a function of a random number in the group for the first element]. Since this is a complex task that involves several phases with individual logical diagrams, an element that is redundant in one phase can be in one or the other row, and sometimes in a different redundant configuration. The fact which applies here is that if it does not appear in one phase—it will not even appear in the following phases. The program must make it possible to draw the condition of each component from one phase to the next, until the simulation of the complete task, while the determination of one successful path per phase is not enough if more than one phase is involved. The condition of each element must be fully determined at each stage. The probabilities of setting aims for solving the complex task of determining the reliability of CTSs can be individually required, while the probabilities of security and success of tasks are most often determined and interconnected. Further decisions defining the continuation of the task depend on the number of available redundant paths for performance of critical functions. At the same time, the probability of successful achievement of other noncritical functions may require a simulation, whereby the computer program must allow simultaneous determination of all these interdependent probabilities.

The boundary method as a reliability calculation method using limit values is applied when the reliability has to be established for the technical system or for the simplest redundant configuration [2]. It is characterized by time savings in relation to considerably longer procedures with mathematical models. It is suitable for the most CTSs in which exact mathematical models cannot be developed, provided that the simulation procedures have to be previously developed. The boundary method involves calculation of the numbers of upper and lower forecasting boundaries, whereby the calculation of the probability of the operational condition or the failure condition and their combination becomes quite simple. Values of a failure event(s) are deducted from the unit (the upper limit of reliability), while the probability of successful cases is added to the unit (lower limit of reliability). By considering as many cases as possible, the area between the upper and lower limits of reliability narrows. The first calculation of the upper limit takes into account only those elements of failure that can individually cause nonperformance of the task, so serially bound blocks in the logical diagram should be considered for the boundary method. This is sufficient only for some satisfactory assessment. In the case where the reliability associated with individual blocks is not very high, for CTSs requiring high reliability in operation, it is necessary to observe parallel blocks.

The Markov process describes the future condition of the system based on the current parameters and thus makes the past and future condition of the system independent [4]. When it is not about continuous but discrete sizes, we can talk about Markov chains [3]. Markov processes are suitable for assessing the reliability of functionally complex systems and complex repairs or maintenance strategies. However, they support the monotony of functions and processes. A model based on the Markov process assumes that the system has the final space of condition and a series of possible transitions between these conditions. Functions, different failure and standby models, and various maintenance activities can be described as different conditions. If the transitions between the conditions can be approximately described by stochastic processes based on the characteristics of this model, then Markov methods can be used to estimate the reliability of the system after several conditions [13]. Thus, it is quite common to use Markov theory to model the problem of predicting the reliability of a repaired system [14]. Markov chains is applied when rates of transition, for example, malfunction or repair depending on the condition of the system, vary depending on the load, level of stress, system structure, etc. Especially, the system structure (e.g., standby conditions) and maintenance policy may create dependencies that cannot be balanced by other techniques. Markov models also have many limitations. They are often applied to repairable systems; however, it is not easy to reach the probabilities of all the transitions that are necessary, and the assumptions of the models are always very restrictive. Also, there are problems in the domain of continuous sizes, for example, Markov processes of mathematical solutions of equations can be very inaccessible, due to which the applicability of the model, as well as in many other cases, is seriously questioned.

The Poisson process was named after French mathematician Siméon Denis Poisson. It is a stochastic process in which events occur continuously and independently of each other. Poison processes are a special case of Markov chains. Poisson processes can describe various phenomena and system failures. This model assumes that the failures are independent of each other and that the number of failures in each time interval is subject to Poisson distribution [15]. There are various types of Poisson processes, such as the homogeneous Poisson process (HPP), the nonhomogeneous Poisson process (NHPP), the complex Poisson process (CPP), the doubly stochastic Poisson process (DPP), the filtered Poisson process (FPP), etc. The HPP as a model requires stationary increments while the NHPP does not require incremental increments. In many applications of the Poisson process, it is not realistic to assume that the average rate of failure is constant. This rate depends on the time, that is, the changeable t. That is why NHPP is a more suitable model for modeling repairable systems with imperfect maintenance actions and a lot of models are based on it today. Also, this model may include rates of occurrence of failures (ROCOF) when they are interdependent, and the time between failures is neither independently nor identically distributed [16]. Some researches suggest that multicomponent repair systems cannot be modeled on continuous distributions, which is logical due to the complexity of the reliability of the system being monitored [17]. Failures which appear in repairable systems could be considered as series of discrete events that appear randomly in the continuum. These situations behave as stochastic points of the process and can be analyzed by statistics of series of events. The log-linear NHPP model and power law NHPP model are two widely used models for repairable systems. Power law NHPP model is based on Weibull distribution. Pulcini applied this model to the reliability of complex repairs. Reliability is software modeled based on the Poisson process and the reliability of software based on this process is also modeled. The Poisson process is suitable for the analysis of repairable systems with more regular failures with the point stochastic nature of the process. However, the existing models on this basis are only available for random failures, but do not include growing hazards over time. Models based on the Poisson process assume that the probability of system failure follows the Poisson distribution, and the number of failures does not affect the reliability of the system. The HPP model assumes that the reliability immediately after the repair is the same as the reliability immediately before the corresponding failure, which makes the model suitable for the so-called minimum repairs, but not for repairs such as overhaul [14]. With the increase of diagnostic techniques, maintenance has got the ambition to improve the constancy of reliability prediction. The proportional risk model (PHM) developed by Cox is currently the most popular model of this type [15]. Similar is the proportional intensity model (PIM), although the first model is more flexible and as such avoids certain problems which appear with the other. Prior to the concept of a proportional hazards model, the reliability and hazard functions were mathematically defined as follows. The reliability function $R(t)$ was used to represent the distribution of the random variable T of the homogeneous population of individuals each of which has a "failure time" [16].

Bayesian methods are brought into connection with Thomas Bayes (1702–1761). There are many of these methods, for example, Bayes factor, Bayesian game, Bayesian multivariate linear regression, Bayesian network, empirical Bayes methods, etc. [10]. The Bayesian model is also used in modeling reliability and it also implies the possibility of implementing empirical experience of maintainers. Mazzuchi and Soyer expanded this model to a traditional time-based replacement policy and a policy of replacing a block with minimal repairs under the assumption that the repair costs are constant and that the parameter of scale α and parameter of shape β are initially independent [17]. Considering the cost of repair of accidental and unknown system failures, Shue developed a model of adaptive replacements using the Bayesian approach, assuming that the hazard of system is monotonous and rising [18]. Using the Bayesian approach, Percy researched the possibilities of improving preventive maintenance [19]. Apelan tried to use a completely subjective or Bayesian approach for making of maintenance decisions when objective data were insufficient [20]. The Bayesian approach is used to describe failures with a common cause. The Bayesian model allows the empirical experience of maintainers to be involved as the biggest advantage but it is not in itself appropriate for modeling the reliability. Its greatest advantage is a combination with the theoretical probability distribution, in which the Bayesian method serves to correct this distribution as the main to describe the failure of the system.

Technological advance makes it possible to use computer techniques to increase analytical capabilities and quality of research [4]. Some of the advanced methods used in the researches are the fuzzy logic, neural networks, genetic algorithms, data fusion, combinations of Monte Carlo methods or Markov chains or these techniques, etc. All these methods are attractive for maintenance since they offer new possibilities in terms of the precision of reliability prediction. Some researchers combined different models (hybrid models), such as the Bayesian method with the Poisson process, or even three models, such as Bayesian, Markov chains, and the Monte Carlo method, resulting in the creation of new specific reliability models [11,21–25]. Hybrid models have a perspective in terms of scientific research, but for now they do not provide universal solutions and have a serious problem at the level of applicability. In recent times, models that relate both to improper maintenance and attempts of prediction of system reliability have been developed. However, most models have serious limitations that limit implementation, but also do not create a realistic picture of the system reliability [10].

1.3 Mathematical Models of the Growth of Reliability of System

The process of system development is a constant interaction between testing and determination of types of failures and changes, which are directed to the elimination of these failures [26–34]. The analysis of the influence of the introduced changes on the system reliability, by applying appropriate mathematical models, is very important for the assessment of the development process. It entails an iterative design-develop process that includes [35] detection of failure modes, identification of root causes, feedback of problems identified, redesign based on failure mode root causes, implementation of redesign, and verification of redesign effectiveness by retesting and iterating the process, Figure 1.2. The majority of models of reliability growth provide the making of conclusions about problems related to the existing system reliability and reliability projection in the next stages of development. The largest number of reliability models examine a certain mathematical formula, which represents the reliability of the system during the development phase. It is commonly assumed that these curves are nondecreasing. When the exact shape of the reliability growth curve is known before the beginning of the development, it is a deterministic model of reliability growth. However, in most cases, the exact shape of the reliability growth curve is not known before the start of the development phase, but it is assumed that the curve belongs to a certain parameter of the curves of reliability growth. The analysis is reduced to the statistical problem of assessment of unknown parameters from experimental data. The assessments can be revised by collecting new data in further development. Assessments are used to dynamically manage the reliability of the system. Concerning the parametric models

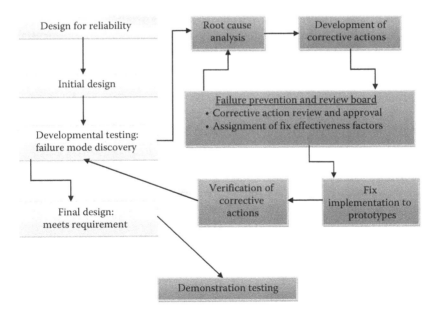

FIGURE 1.2
Reliability growth testing process. (Based on the Page 10. "AMSAA Design for Reliability Handbook" [35].)

of reliability growth, Duane's and Wieren's models are emphasized, and concerning the nonparametric models, the Barlow and Scheuer models.

Duane's mathematical model for describing the intensity of the failure is as follows:

$$L(T) = \beta T^{-\alpha}, \text{ for } 0 \leq \alpha \leq 1 \text{ and } \beta > 0, \quad (1.1)$$

where $L(T)$ - cumulative intensity of the failure during the working time T, α and β - parameters.

Duane's model is very important for the analysis of reliability growth and is characterized by the following advantages: the model is mathematically simple and, in practice, it is often applied; since the parameter α is of a dimensionless size, it is relatively easy to establish dependency on the level of effort involved, it is possible to use cumulative data for tests. The similarity of Duane's model to the Weibull model allows the use of existing analytical methods for reliability intervals and statistical testing of the hypothesis; the model is represented by the straight line on the log-log paper of probability, which is convenient for illustrating the reliability growth curve. The limitations of Duane's model are as follows: the assumption of the model is that the failures belong to the exponential distribution, although the attributive tests can be analyzed if Poisson approximation for the binomial distribution is used; the model assumes that the corrective actions are carried out simultaneously with continuous tests, although it is successfully applied even when testing and corrective actions are interchangeable and when there is enough time to make changes to the system which is examined; if it happens that the form of growth of reliability cannot be precisely defined by the basic Duane's model, then the whole development period is divided into a number of parts, so a special Duane's model is applied for each part.

Wieren's model of reliability growth. The mathematical model of Wieren's growth of reliability can be described by the equation:

$$R = ab^c, \quad (1.2)$$

where the parameter a represents an unknown upper limit of reliability R, which is asymptotically achieved when the development time $t \to \infty$, while $0 < b < 1$ and $0 < c < 1$, parameters which are assessed based on the test results. At the moment, $t = 0$, the reliability level is ab. Therefore, the parameter c determines the form of growth of reliability. Wieren's reliability model implies the assessment of parameters a, b, and c, while a certain procedure is required.

Reliability model of Barlow and Scheuer. This model assumes that the changes during development do not decrease the reliability of the system, but no functional form for reliability growth is defined. In this model, each failure must be classified as an inherited failure, in which the causes of the occurrence cannot be determined, or as a causal failure, the causes of which can be determined. The development program takes place in K phases, while similar systems are tested within each phase. It is considered that one phase of development is completed when the causal failure occurs, except for the last one. At each stage of development, the number of systems whose tests have been successfully completed is recorded. According to this model, it is assumed that the probability of inherited failure of q_0 remains the same during the whole development phase, and that the probability of causal failure at q_i at i stage does not grow from one phase to another within the development program, that is, must be $q_1 \geq q_2 \geq \cdots \geq q_K$. In order to eliminate the cause of failure, certain reconstructions are performed, so that the tested systems differ from phase to phase, but are homogeneous within a single phase. Reliability models include some assumptions, so the question is how much they are acceptable for presenting the actual process of reliability growth. It is important that the availability of appropriate statistical procedures necessary for the assessment of the relevant characteristics must be taken into account. However, these models provide a mathematical description of the empirical and planned reliability growth, as well as the design of reliability for the next stages of development. When the characteristics of reliability growth are poorly known, nonparametric methods are used, in which the lower limit of confidence for reliability is very low. Reliability models are used to consider changes in system reliability during the program of development. Regarding the time period of use and the goals to be achieved, there are four models of reliability growth: empirical growth model, planned growth model, model of growth assessment, and the model of the projected assessment of growth.

Empirical reliability model provides an answer to the question of what kind of form of reliability can be expected. The model provides a general form of growth and is based on data from the development of similar systems. It can be graphically represented by showing some reliability characteristics (mean time between failures, probability of success, etc.). The most common form of the empirical model is given by a continuous curve, whose equation is determined by one of the mathematical models of reliability growth (often, Duane's model). The continuous curve represents growth in the phase or part of the phase, where changes are made in the design of the system being tested and where the growth of reliability is realized in a series of jumpy changes. These jumpy changes can be negative, which is often the case at the beginning of the production process, when production is still in the process of establishment.

The planned reliability growth model is used to solve the question of what values are expected in certain periods of the development phase. The curve of the planned growth model has the same general form as the empirical curve, but passes through a determined set of points, which are determined from the characteristic phase. These determined values depend on the complexity of the system, time of testing, failure analysis, and possibilities for reconstruction. The initial reliability values are mainly determined by the synthesis of the test results of the elements and subsystems, or based on the results of prototype testing. If the initial test results in the development of the system do not match the predicted initial reliability values, the program must be carefully reexamined to assess the forecasting reality.

The reliability estimation model provides an answer to the question of the extent of existing reliability at a certain point in time. This assessment can be achieved with three possibilities: based on the results of testing of the existing system project; based on a statistical combination of the results of all previous tests, taking into account the achieved growth of reliability; based on the results of all previous tests, with a preliminary assessment of the results of previous tests. The model of the projected assessment of the growth of reliability provides an answer to the question of what is expected at the end of the development program, if certain planned activities are followed. The projected assessment can be obtained by extrapolation starting from the currently assessed value, using the empirical model for determining the general shape and the proposed program characteristics for determining the specific direction of further development.

1.4 Reliability Assessment Methods

An important step within the security analysis, and therefore the reliability of technical systems, is simply the standardization of security, that is, the formulation of the requirements for the security of system. In addition, the problem of formation of the minimally sufficient set of indicators, which characterize the observed property of certain system, is still not completely solved [36]. Depending on the observed system, the security, or reliability as its component, is the result of superposition of other more "elementary properties," such as mechanical strength, stability, refractoriness, elasticity, etc. The existence of potential sources of danger and thus the rate of hypothetical failures can serve as a universal quantitative feature of safety or reliability of all technical systems. Thus, through this indicator, it is possible to mutually compare the technical subsystems of different purposes and operating principles, that is, "measurement" according to the scale of emergencies of different sources of danger. This represents a risk, which is characterized by the frequency of occurrence of unwanted events in the unit of time. In the dictionary of the European Quality Organization (EOQ), within the terms used in the field of general quality management, the risk is defined as "the common factor of probability of occurrence of unwanted events and their consequences" [37]. Until recently, the basic methods of analysis of reliability as a component of the broader term of security were based on the conservative concept of "absolute security," which is not adequate to the probable nature of occurrence of failures and disorders of exploitation, caused most frequently by changing the conditions of exploitation. On the other hand, in order to avoid the occurrence of common differences between the set requirements for reliability and their dependency on the fulfillment of operational requirements, special attention should be paid to defining the analytical expressions and numerical values of reliability parameters [38]. For the realization of this task it is necessary to form an appropriate database, related not only to the system as a whole, but also to the components of the system as the basic links in the reliability chain. The intensity of failure of some of the system components depends on many factors (mechanical and thermal overload, environmental impact, exploitation conditions, the manner of repair or replacement, the influence of human factors, etc.). In addition, the reliability assessment, depending on the purpose and phase of the life cycle of the thermal power plant, is generally realized in three basic manners: estimation of reliability on the principle of similarity of equipment, on the basis of its typization or retrospective analogue information, with the correction for new forecasting project conditions, then the reliability assessment with the method of listing components, or so-called rough reliability calculation, with the formation of appropriate statistical methods and logical-credential models, as well as assessment in case of incomplete determination of information and reliability assessment by the stress analysis method, or the so-called fine calculation reliability (characteristics of possible relations of work parameters and loads), as well as assessment of probability of endurance parameters and possible deviations of constructive elements, expert correction of characteristics of durability and resources of details with the participation of harmful impacts [39,40].

The intensive development of probability methods of security analysis resulted in the formulation of a set of probability methods for analyzing the security of technical systems [8]. Further progress in improving the reliability assessment, except in the adaptation of classical methods to the specificities of the complex of the thermal power plant, lies in the need to shorten the testing time of one or more factors by selecting an optimum plan of shortened testing by automating on-line procedures of reliability assessment and its optimization based on selected criteria (most often the economic criteria) [41]. Taking into account the very structure of the thermal power plant technological system and the characteristics of reliability of certain elements, it is also necessary to provide the measure of importance and rank through it the elements from the aspects of rational distribution of resources while increasing the reliability of each of them. As a result of solving the problem, the list of the critical consequences of the consequences (effects) of the failure is determined. The conditions necessary to be fulfilled in order to get the list are the following: knowledge of the conditions of operation of the thermal power plant system, its structure and possession of the database of failure of elements. Setting the methodology and criteria, based on which the priority lists for replacements and reconstructions of individual units within the CTS of Co-TPP would be determined, results in the maintenance of satisfactory safety of the entire Co-TPP system as a whole and reduction of the operating costs of the electrical energy system (EES) as a higher hierarchical system [42]. The "weak points" in the complex system of the thermal power plant significantly affect the reliability and safety of their work and carry certain risks. An increase in the level of security and reliability of any system element is directly reflected in the system as a whole. All this requires that the considerable attention has to be paid to the issues of reliability and availability of individual components and the system as a whole. Starting from the assumption that the manners of providing reliability at different stages of lifetime can be provided through taking certain types of reserves, with the elimination of all other redundant parameters, both for the basic and the extended lifetime, the following are stated as their most frequent representatives: functional form of reserve, reserve in load, time reserve, assessment of the dependency of the reliability of plant on the adopted programs of overhaul and the contents of overhauls, and the assessments of reliability, security, and durability of the thermal power plant with the participation of the technological and information form of the reserve.

1.5 Indicators of Reliability of Co-TPP

The issue of reliability of operation of Co-TPP is of special interest, given that about 70% of total electricity production is realized on these plants. Since it is not possible to have a stock of electricity in the warehouse, it is necessary to have an adequate reserve of installed power during the operation of the Co-TPP within the EES, and any change in the consumers' demands for electricity determines the production of electricity within the EES [43]. The work of any electric power facility is strictly defined by legislation, rules, and instructions, on the basis of which the operating conditions and safety of their exploitation are defined. Quality, reliability, safety, economy, and ecology are particularly important for the operation of electric power facilities. The reliability is the probability of realization of the required function of the objective of energy system (production of electric and heat energy and/or technological steam) in the given scope and concrete conditions of exploitation, Figure 1.3.

The energy block within the electric power system works with the adequate installed power N (set 1) according to the specified graph of load. The probability of such operating mode in specific conditions is given with P_N (set 2) in time duration τ_{rad} (set 3) [44]. The reliability of the energy block is determined by its operating ability for production (set 4) in the form

$$E = NP_N \tau_{rad} \qquad (1.3)$$

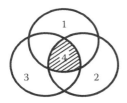

Legend:

1 – a set of given objective functions;
2 – a set of given operating conditions;
3 – a set of time intervals;
4 – a set of data, characterized by reliability.

FIGURE 1.3
Illustration of the concept of reliability of the energy system for the production of electricity and heat and/or technological steam. (Based on Milovanović [44].)

which represents the state of the energy facility in which it is able to fulfill all or part of the given function (work with reduced power or production of one form of useful form of energy) in the required extent. The loss of operation ability is a failure of the system. In the process of exploitation, there are cases when there is a complete or partial loss of functional properties. The event which results in the discontinuance of the energy system operation is called *failure*. Failure can be *complete* (emergency interruption or downtime) or *partial* (reduction of work ability), Figure 1.4. Failures can be *immediate* or *gradual*. Immediate failure is characterized most often by breakage and devastation of individual elements or parts of the power system, which automatically by its function imply its full downtime, while a gradual failure has

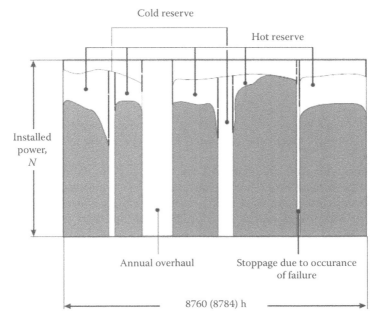

FIGURE 1.4
Overview of operation of power facility within EES [2]. (Based on Milovanović [37].)

a time change in the state of one or more plant elements. These failures have been caused most often by the weakening of the material due to work in thermally unfavorable conditions or caused by removal of material and the reduction of walls due to corrosion, erosion, and abrasion.

In addition to the criteria for assessing reliability indicators, it is also necessary to define the basic and *supplemental reliability indicators*. The choice of basic and supplemental reliability indicators is directly related to the conditions of specific tasks [44]. There are different indicators in the stage of development and design of facility, the solution of the tasks of optimization of the power system and its components, in the stage of production of serial energy equipment and details, in the stages of installation and commissioning, as well as on the stage of exploitation. Failure and stages of recovery of operating ability represent opposite events that make up the flow (sequence) of events, Figure 1.5.

The safety of functioning of the energy system and its accompanying energy equipment is determined by the number of different (by their nature) factors, such as construction, quality of materials used, production technology, quality of installation, service and exploitation conditions, quality of steam and similar. The flow (sequence) of events can be described by the sequences of distribution of random sizes, which characterize the probability of occurrence of these events $P(k)$, where k represents the number of failures (random events). Thus, the probability of event X

$$P(X) = m^*/n, \tag{1.4}$$

where m^* is the number of random events and n is the number of all events.

The probability of work without failures for the repaired (newly recovered) to the planned operation time T_0 is determined as follows:

$$P(\tau) = e^{-(\lambda/\tau)}, \tag{1.5}$$

where τ is the observed time interval and $\lambda = 1/T_0$ is the intensity of failure, Figure 1.6.

The parameter that characterizes the frequency of failure for a certain period is the *parameter of flow of failure*, and represents the rate of the probability of occurrence of failure of the object which is repaired, or the average number of failures of equipment which is repaired in the unit of time.

FIGURE 1.5
The course of failure and its removal with the aim of restoring work ability ($\tau_1, \tau_2, \ldots, \tau_n$—operation time until the failure (time from the start of operation until the occurrence of failure; $\tau_{rep1}, \tau_{rep2}, \ldots, \tau_{repn}$—time of repair). (Based on Milovanović et al. [44].)

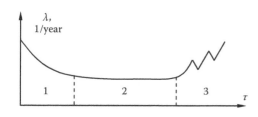

FIGURE 1.6
Intensity of failure during service life of elements. (Based on Papić [1].)

Legend:
1 – area of frequent failures (early failures);
2 – area of normal exploitation (random failures);
3 – year cancellation due to the aging of the equipment (late or delayed failures)

The reliability of the facility of the energy system according to its function, that is, the giving of mechanical work through the connection to the generator and the production of electrical as well as thermal (and technological) energy according to a predetermined strict regime with regulated and unregulated deductions, can be characterized by appropriate complex indicators, of which the most significant is so-called *coefficient of production of insurance (power, energy)* [1,11,44].

Table 1.1 gives an overview of the planned operation time to the failure T_0 and time of repair T_{pop} for 200, 300, and 800 MW power blocks and their most important elements (boiler and turbine plant). As specially performed cases of coefficient π, in the literature there is often also a *coefficient of technical exploitation* K_{ti} and a *coefficient of readiness* K_g. The coefficient of readiness is characterized by the probability of a state of work ability at an arbitrarily selected time for the element characterized by alternative conditions "work (exploitation)—repair (recovery)," calculated on the basis of the following equation:

$$K_g = T_0/(T_0 + T_{rep}) = \mu/(\lambda + \mu), \qquad (1.6)$$

where $\lambda = 1/T_0$ and $\mu = 1/T_{rep}$ are appropriate intensities of failure and repair.

TABLE 1.1
Overview of the Planned Operation Time to the Failure T_0 and Time of Repair T_{rep}, h

Energy plant	Planned operation time to the failure T_0 (h)	Time of repair or recovery T_{rep} (h)
Block K-200-130	800–1000	45
Boiler aggregate	900–1100	55
Turbine	5000–6000	30
Block K-300-240	800–1000	43
Boiler aggregate	300–500	60
Turbine	4000–5000	90
Block K-800-240	600–800	80
Boiler aggregate	900–1100	90
Turbine	3000–4000	60

Source: Data from Milovanović [44].

The frequency of failure of an element is evaluated by the number of damage resulting in the exit from the plant in the unit of time and is determined as the ratio of the number of elements with the failure for the period $\Delta \tau$ compared to the total number of one-type equipment, that is, it is applicable:

$$\omega = n_0/(n\,\Delta\tau) = 8760(8784)/T_0 = 8760(8784)\lambda. \tag{1.7}$$

The time of repair of the element is determined as the time of overhaul increased by the duration of the diagnosis for the purpose of finding the defect.

The technical exploitation coefficient represents the ratio of the expected value of the time during which the facility was in operational condition for a period of exploitation and expected values of the total working state of the equipment of steam turbine, technical maintenance, and overhaul time. It should also be noted that the coefficient of readiness is a probability that the individual equipment and turbine as a whole will be ready to work at any given moment, except for the planned periods for performing planned overhauls and technical maintenance tasks.

Elements can be connected within the energy system either serially or in parallel or combined in series and parallel. An example of a serial connection is the connection of elements within the main power facility (MPF) of a thermal power plant (TPP), where the failure of any of the three elements means a failure of the system as a whole, Figure 1.7. The following equations apply to this connection) [44]:

$$\omega = \Sigma \omega_i, \tag{1.8}$$

$$t_{rep} = \Sigma \omega_i T_{repi} / \Sigma \omega_i. \tag{1.9}$$

The parallel connection of the elements is characteristic for boiler plants connected to the common collector, from which fresh steam is supplied to the turbine plants or to the facilities that are reserved within the higher hierarchical EES. In energetics, for a quantitative assessment of reliability, a number of complex indicators are used, from which the factor of power of block, factor of utilization of block capacity, factor of exploitation, factor

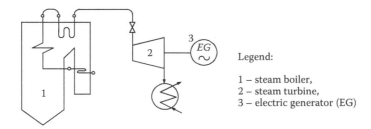

FIGURE 1.7
Presentation of serial connection of elements of MPF to TPP. (Based on Milovanović [36].)

of stoppage, etc. are used. The factor of power of block K_R is defined as the amount of realized and nominal (computational) power, and is calculated according to the formula $K_R = R_0/R_N$, where R_0 is the achieved mean power in exploitation, and R_N is the nominal power (for new blocks the power factor is 1 and for the elderly 0.95–0.98, where deviations to 5% are allowed).

The factor of utilization of capacity of the block K_{ik}, which is determined based on the generated electricity and is defined as the quotient between the amount of generated electricity and the theoretically maximum possible generation with 100% used capacities of the facility of the block of the thermal power plant, is calculated according to the formula $K_{ik} = E_0/E_m$, where the generated electricity of the block of the thermal power plant is $E_0 = R_0 T_e$, that is, $E_m = R_N T_k$, theoretically maximum possible generation of the block of the thermal power plant. The following parameters are applicable: R_0—the realized mean power in exploitation in MW, T_e—the realized exploitation time in hours, R_N—the calculated nominal power in MW and T_k—the calendar time in hours.

Next (it is applicable) [44]:

$$K_{ik} = R_0 T_e / R_n T_k. \tag{1.10}$$

Since the power factor is $K_R = R_0/R_N$ and the exploitation factor is $K_E = T_e/T_k$, the capacity utilization factor of the block is obtained:

$$K_{ik} = K_R K_E. \tag{1.11}$$

This coefficient is in the range of 0.40–0.80 depending on the age of the block, its performances, and operational readiness. The exploitation factor is one of the indicators of the time exploitation of the block of the thermal power plant and can be defined in two manners: the most commonly used in literature is the definition of the block through exploitation time—the work of the block on the network (T_e) and the calendar time (T_k) for the observed period (monthly, annually, etc.), that is, is $K_e = T_e/T_k$. For the new block, the annual exploitation factor is 0.8, while for the old blocks it ranges from 0.6 to 0.75. In this way, a certain coefficient of exploitation is used in analyses, since the overhaul time is not excluded from the calendar time, which is important in comparative analyses. The other manner for determination of the annual exploitation coefficient is through the time of exploitation and the difference between the calendar time and the time determined for the annual overhaul, $K_e^1 = T_e/(T_k - T_r)$. This manner of calculation is more used at the local level and in the preparation of annual plans. In most thermal power plants, the duration of overhauls is not standardized, and for a detailed analysis, the other manner of determination is not reliable for the evaluation and comparison of the quality of the operation of the thermal power plant.

Stoppages of thermal power plants are caused due to disturbance in the technological process resulting from nonstationary operating modes caused by the failures of the plants and the vital parts of plants. The failures which condition the stoppage violate the secure and safe operation of plants by the impossibility of maintenance of technological parameters according to technological instructions and technical regulations of exploitation. Some failures, which require urgent termination of the power plant block, may endanger the safety and security of personnel at exploitation and maintenance. That is the reason of existence of protection and blockages for immediate disconnections after the occurrence of such cases. Failures of plants and equipment of the thermal power plant block which condition the stoppage and shutdown due to recovery can be classified into two basic groups: the *emergency* ones, due to which protections automatically act and exclude the block and the *nonemergency* ones, because of which the work for some time can be extended and technical regulations on exploitation and security and environmental safety will not be endangered. That is why the stoppages are grouped into two basic groups—unplanned and planned stoppages. Unplanned stoppages happen as a result of failure in operation of the block caused by abrasion of the material, aging and loss of functional properties, thermal overload, improper exploitation and maintenance failure, noncompliance with technical instructions and regulations, inadequate overhaul, weariness, deteriorated quality of basic fuels, etc. Since the thermal power blocks are complex units, and are made up of a large number of dependent technological units and complex plants, thereby the possibilities for unplanned stoppages are higher, especially if the blocks are older. Planned stoppages include annual overhauls and overhauls for care of plant. Analysis of quality of exploitation and maintenance can be carried out through the thermal power plant failure factor. Factors of stoppage include: failure factor, factor of planned stoppages (care), factor of overhaul, and factor of suppression from the network.

The factor of failure or exclusion is defined as the quotient of time duration of removal of failure due to which the outage or exclusion of block from the network happened immediately after its occurrence and the calendar time for the observed time interval. It is most often calculated at annual level, and can be calculated for other requested time period of exploitation of thermal power plant. It is calculated according to the formula $K_{KV} = T_{KV}/T_K$, where T_{KV} is the time duration of the shutdown due to the failure in hours or exclusion in the observed time interval. In fact, this is the time period from exclusion of the block from the network to re-inclusion or synchronization to the network. If there are more stoppages, then these times are aggregated depending on the period for which the analysis are done, that is, applies to

$$T_{KV} = T_{KV1} + T_{KV2} + \cdots + T_{KVn}. \tag{1.12}$$

The value of this factor is in the range 0.1–0.15, calculated on an annual basis. Lower values are for new blocks and larger for old blocks. Planned shutdowns are introduced as preventive measures in order to achieve greater operational readiness and risk reduction, and prevent emergency cases in the work of the blocks of thermal power plants. The size is important while planning the investments in older blocks or solving the problematic cases in exploitation that do not endanger safe work, and they need to be annulled in a certain period of time.

For the planning of shutdowns of old blocks, which are at the end of their service life, the factor of planned shutdowns is the most commonly used, whose values are most often defined on the basis of monitoring of the exploitation data of the thermal power plant and similar thermal power plants. It is defined as the quotient of the time duration of the planned shutdown and the calendar time for the observed time interval. It is calculated according to the formula $K_{PZ} = T_{PZ}/T_K$, where T_{PZ} gives the time duration in hours of the planned shutdowns in the observed time interval, and it is defined from the disconnection from the network to the re-synchronization. In fact, this is the time period for which the failure has been removed due to which the block has been suspended or the planned investment has been made on the block of the thermal power plant or the sum of more such time periods, if any, during the year. It is most often determined at annual level and is rarely applied in planning. It is used in special cases with older blocks and larger investments, and at the end of the service life of thermal power plants.

The overhaul factor defines current and capital annual overhauls. It is defined as the quotient of the overhaul duration and calendar time at the level of the year. It is calculated according to the formula $K_R = T_R/T_K$, where T_R is the duration of the overhaul in hours, that is, the stoppage of block for performance of planed overhaul, and is defined as the exclusion from the network to the re-synchronization. The overhaul factor is determined at the year level and has different values depending on whether the blocks of thermoelectric power plants are older, of higher power and the extent of planned overhaul work (whether it is current or capital annual overhaul). The values of this factor for older blocks range from 0.08 to 0.16. Higher values refer to capital, and less to current overhauls.

In the practice of exploitation of the thermal power plant blocks, there are cases when the power system network due to reduced consumption or for some other reasons, the production increase which is not covered by the consumption. In that case, the dispatching service must maintain the balance of the system and exclude the individual blocks from the plant, that is, suppress the block of the thermal power plant from the grid. This occurs most frequently when the water supply in hydroelectric power plants is uncontrollable large or in cases where the reductions in consumption are unplanned, that is, there is no known electricity purchaser. These cases are rare, but they happen in practice. They are defined by factors of suppression and are not of planned size. The suppression factor is defined as the amount of time of depression and calendar time. It is determined at the level of the year by the formula $K_P = T_P/T_K$, where T_P is the duration of the suppression in hours, that is the blockage of the block from the exclusion from the network to the re-synchronization, and $T_K(h)$—calendar time on an annual level.

1.6 Failures and Damages during the Operation of Co-TPP

The failure of a part of the thermal power station system or the thermal power station as a whole is defined as the termination of the possibility of some system element or the system as a whole to perform the functions they have been designed for [11]. The reduction or the loss of the technical system working capacity in the course of exploitation is a consequence of the effect of various factors (embedded, random or time), which change initial system parameters, causing alongside also a different level of damage. For reducing unplanned jams, preventing breakdowns, and increasing reliability in the work of the individual parts of thermal power stations or the thermal power station as a whole, it is necessary to strictly apply the regulations for quality insurance in the course of their lifetime, starting from the phase of preparation and design and all the way to the end of exploitation and its withdrawal from the operation. During the exploitation period, the degradation of the condition of both the elements and the thermal power station as a whole is necessary. Monitoring of the condition in a specified time period or continuous monitoring represents a process of constant inspections or supervisions of the equipment operation for the purpose of ensuring a proper functioning and detecting the abnormalities that announce a forthcoming failure. It is suitable for the equipment for which it is not possible to predict the wear-out trend by the periodic checks. The very modeling of the system conduct most frequently depends on the specific operational research, application of the mathematical statistics method (definition and selection of distribution, assessment of observed parameters, hypothesis test, definition of the scope, and estimation of characteristics), as well as on the application of the probability theory method (different mathematical models) [38]. The functions that the technical diagnostic system should realize are given in the form of specific checks of the system technical condition, checks of the working capacity, checks of the functionality, location of the failure position at the lowest possible hierarchical level, as well as estimation of the remaining period of use or trend of the malfunction occurrence, Figure 1.8. The application of technical diagnostics opens up new possibilities of managing electrical power stations, which creates all preconditions for a significant decrease of corrective and preventive activities related to maintenance, along with preserving the same or realizing a higher level of reliability of the plant as a whole. By introducing maintenance according to the condition, along with the application of the technical diagnostics and proper determination of the remaining operational lifetime (reliability management), it is possible to decrease the number of failures of the system of the steam turbine and electrical generator) [1]. Of course, this has to be followed also by the application of the computer technique, as well as the database both at the level of the thermal power stations and at the power utility level. Besides, the diagnostics enables a good quality assessment of the aging progression

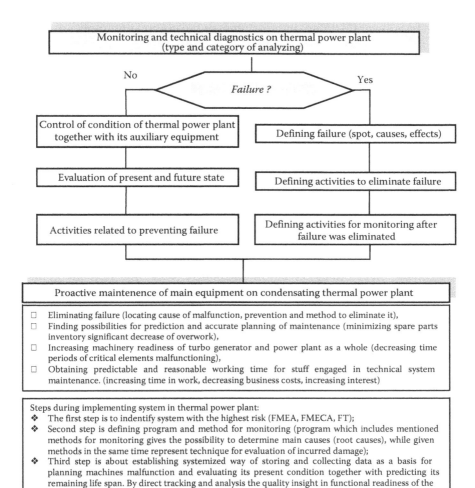

FIGURE 1.8
Functions of technical diagnostic and monitoring related to condition of certain parts of thermal power plant. (Based on Milovanović [38].)

and the remaining lifetime, and specification and planning of the restoration, the replacement of the old equipment, as well as optimal correction, that is, it is tightly related to the maintenance strategy according to the equipment condition, which makes it directly affect reduction of the costs originating from the cuts in production, transmission and distribution of the electric power [37]. The supervision over the equipment implies an automated and continuous determination of its status, along with following the values of several parameters within the plant. Depending on the number and type of the controlled parameters, we differ partial (following one or several related values) and complete monitoring systems (following a hundred different parameters of a certain plant element). Here, it is important to mention that the complete monitoring systems often also contain the expert subsystem, which based on the collected data and the diagnostics relying upon the embedded expert knowledge and algorithms at an early stage warns the operator about the forthcoming problems and recommends necessary actions. It is not difficult to show that the purposes of diagnostics and monitoring are identical—increase of the plant cost-effectiveness and equipment availability. We can say that the automated diagnostics "with no time tensions" is in fact a synonym for the system monitoring [38,46].

The most common classification of failures in TPP includes their grouping into the failures due to structural defects (defects of technical documentation, errors in calculations and mathematical modeling, wrongly applied methods of calculations, etc.) and low quality of production, errors in exploitation (noncompliance with the operating mode of EES, noncompliance with production guidelines and instructions, accidental mistakes of the workers), low quality of assembly work, and defects of overhaul. Design and assembly errors are detected in the period up to 30,000 h of operation. Figure 1.9 shows the distribution of the failures in TEP. Physical and chemical processes in the steam boiler during exploitation are the most complex (steam tract, smoke tract, material of certain elements of the steam boiler), and result in a change in the properties and characteristics of the material. The processes of combustion, heat exchange, corrosion, and formation of deposits on the heat exchanger surfaces determine the reliability of boilers to a great extent.

Characteristic failures caused due to design defects on boilers result in large heat deformations on the heating surfaces caused by the high speed of ash consumption. The distortions of the characteristics of elasticity, casting,

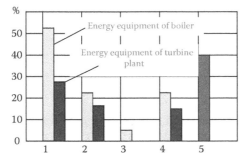

FIGURE 1.9
Distribution of failure on TPP as a whole. (Based on Milovanović [44].)

with thermal treatment of heat-resistant parts made of steel, welded parts, etc., are widespread. The deviation of real charcoal characteristics from the calculated ones leads to deviation from the given volumes of combustion products and temperature of flue gas at the boiler output, and the consequence is the disturbance in the work of the boiler's convective part, the increase in ash extraction. The low quality of steam and water leads to a sudden increase in deposits, with the rise of the temperature of the steel tubes and their overheating.

Table 1.2 gives the intervals for distribution of failures of boiler, depending on their capacity. The intensity of failures of the energy equipment of the boiler plant is not the same, Table 1.3.

During the exploitation, the pipe screens are exposed to the effect of radiation energy, the corrosive active environment of the fuel combustion products, which, at low circulation speeds and disturbance in the water regime of the boiler, results in their damage and failure in the operation of the boiler. It should be noted that the quality of water and steam has a decisive influence on the damages of the heating surfaces of the steam boiler.

TABLE 1.2

Distribution of Failure on Steam Boiler

Boiler Capacity (t/h)	Share of Failures Due to Damages (%)				
	EK	ES	SH	SIS	OBE
2500–2650	40–45	18–21	24–28	6–10	1–5
1600–1800	3.7–4.3	8–11	35–40	45–50	0.5–1.0
950–1000	10–13	20–25	46–51	10–15	4–8
640–670	23–26	14–18	40–46	10–15	5–8
480–500	29–32	21–26	35–40	–	5–10
320–420	27–31	13–16	44–48	–	8–12
120–220	30–34	18–22	38–42	–	5–10

EC, economizer; ES, evaporator surface; SH, steam heater; SIS, steam inter-superheater; OBE, other boiler elements.
Source: Data from Milovanović [44].

TABLE 1.3

Share of Failure of Power Equipment of Boiler Plant

Mark of the Part of Equipment	Share of Failures (%)
Heating surfaces	77–81
Auxiliary boiler equipment	2–5
Fuel supply	1–3
Reinforcement	3–6
Automation	5–10
Other boiler elements	2–4

Source: Data from Milovanović [44].

An uneven field of temperatures along the height of the smoke duct, which has a steam superheater (the heat load of the upper and lower parts of the coils can vary up to 20% and the smoke channel width up to 30%), has significant influences on the change in thermal deformations. Steam heaters are also damaged by prolonged operation at temperatures higher than 500°C, where the metal structure suffers significant changes. The curves on most of the pipelines are damaged due to corrosion-fatigue processes, and due to the lack of compensation for uninterrupted thermal expansion. The main damage to the stop and the control valves are defects in valve housings and valves, deterioration of density, and similar. Compared to the boilers, the failures during the operation of turbines are significantly less frequent. However, physical and chemical processes leading to the reduction in reliability levels of turbine components have much in common with the processes occurring on boiler elements (change of properties of metal over long service life, erosion processes, etc.). Emergency situations happen in the case of breakage of blades, then in the failures in the automatic control system, as well as during damage to the bearings (increased vibration). They are caused by the imperfection of technology of commissioning, disconnection from the drive, and unloading. Damaged blades due to the action of the flow of wet steam are characteristic for the last levels of the part of low pressure of turbine. Rotor damage happens most often due to insufficient quality of manufacturing and disturbance in the operating mode of commissioning and exclusion from the drive, which can lead to the occurrence of residual deviation. Figure 1.10 shows the characteristic distribution of the turbine failure.

The number of failures can be prevented by the application of organizational and technical measures (by ensuring that the observed plant works only on the project fuel, by selecting the optimal operating regimes, carrying out measures and maintenance activities, etc.). Other failures can only be prevented by timely replacement of equipment or some of its elements

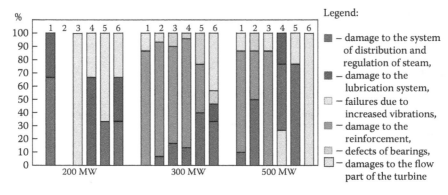

FIGURE 1.10
Overview of main causes of failure at turbine plants from 200 to 500 MW per years (1, ..., 6—years). (Based on Milovanović [44].)

TABLE 1.4

Orientation Indicators of Reliability of Energy Blocks by Years after 200,000 h of Work

Power of Block (MW)	Year	ω(1/year)	T_v(h)	Coefficient of Readiness(%)
200	1	6–8	40–42	97–98
	2	4.5–5.0	30–32	98–99
	3	9–10	37–39	95–97
	4	6–7	36–38	97–98
	5	4.5–5.0	93–97	94–96
	6	12–15	80–84	88–89
300	1	4–4.4	15–17	99–99.5
	2	7–7.4	16–20	98–98.5
	3	6.5–7.0	26–30	97.5–98
	4	6.7–7.2	24–26	97–99
	5	8.5–9.2	50–55	94.5–95
	6	9.0–100	55–60	94–94.5
500	1	16–17	41–43	92–93
	2	20–21	39–41	91–92
	3	26–27	45–47	87–89
	4	20–21	116–120	78–79
	5	18–19	78–80	85–86
	6	14–15	120–125	83–84

Source: Data from Milovanović [44].

(preventive maintenance). Timely overhauls of the high technical level with the participation of normative technical documentation and diagnostic methods provide reliable long-lasting operation of the equipment, Table 1.4.

1.7 Modified Method for Assessment of the Optimal Reliability of Co-TPP

Possible criteria for the selection and formulation of the contents of mathematical models and methods for calculating the reliability of the thermal power systems during the life cycle, depending on the choice of its principled scheme, variants of construction and parameters, methods and character of the predicting the reserves, system for overhaul-technical maintenance, diagnostics, and protection, are as follows: forms of connections in terms of reliability, operating modes of basic and auxiliary equipment, as well as other conditions defining the manners for ensuring the reliability of facilities as a whole and its constituents; the procedure of processing the condition of the working and nonworking activities of the elements and the system as a whole, their interconnection and possible forms of representation of their change in time; defining criteria, basic and additional reliability indicators for solving optimization tasks by elements, and the system of thermal power plant as a whole; defining the limitations and additional requirements of the tasks for assessing the optimal reliability, as well as the additional conditions, the prescribed norm or possible forms of their presentation; scope and characteristics of the initial information (parameters), with the assessment of their completeness, presentation form, accuracy, etc.; the possibility of applying existing programs of computer technology, volume, periodicity, calculation speed, and limitations of existing methods. Analyses of the complex system of the thermal power plant from the aspect of expected reliability and preventive engineering have the task not only to find and remove the "bad spots" in the plant, but also to assess the moment and justification of its revitalization. A timely decision on revitalization will result in the appropriate reconstruction and modernization of both the plant and the system as a whole. In this case, the appropriate economic savings will be achieved in the work of the power plant, and the funds invested will be returned through increasing the reliability indicators, that is, increasing the time in operation, and reducing the time of failures. The algorithm of the modified method for assessing the optimum reliability, shown in Figure 1.11, based on the system of technical diagnostics and condition-based maintenance, will significantly improve the procedure for making such a decision in terms of larger unification of the method on blocks whose nominal power is different from the referent. The starting database was developed as a result of many years of research carried out for the basic configuration of the thermal power plant for solid fuel (coal), with a nominal reference power of 300 MW. Due to the specificity of the work of the considered systems of different thermal power plants, the lower minimum values of the interval estimation of the reliability characteristics of the basic referent block are defined. Any change of a particular project-forecasting condition requires additional consideration of other forms of information sources, mainly from lower hierarchical levels and a greater degree of details of the objects and processes that occur there. It is also necessary to complete the basic method of structural calculation for the reliability indicators for the working and nonworking condition of work with the corresponding additional correction factors (two-step correction system). The rapid and efficient applicability of this method on a range of energy blocks with different installed nominal power, the position within the power system and a specific maintenance system, is enabled by the use of a simple empirical equation [11,43]:

$$C_A = (A/300)^m B_{300}, \tag{1.13}$$

where A is the nominal power of the thermal power plant, MW; B_{300} is the indicator of reliability for a 300 MW thermal power plant system; C_A is the an indicator of the reliability of a thermal power plant with a power different than 300 MW; m is the value of the exponents obtained on the basis

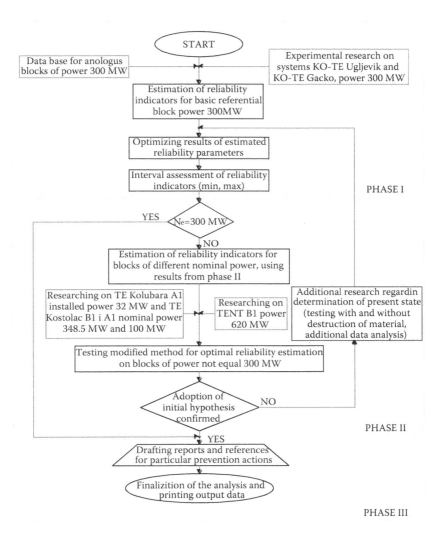

FIGURE 1.11
Block diagram of modified method for reliability evaluation of condensation thermal electric power plant. (Based on Milovanović [11].)

of statistical data processing from exploitation during the life cycle of the thermal power plant.

The values obtained represent the estimated value of the reliability indicator, the accuracy of which is further improved by the iterative process to a certain, previously defined, level of accuracy. For simplicity, the initial use of data in the form of diagrams of analogue objects (the first iteration) is recommended, and, then in the next phase, it is necessary to make the necessary corrections from the aspect of participation of functional and reservation of, as well as participation of time reservation, etc. We should not forget the correction of the assessment of dependency of reliability indicators in relation to the overhaul programs, as well as the contents of the overhaul cycles. At the level of development and design, previous calculation comparisons of the mono and double block structure, as well as the parameters of the selection of operational stocks, fuel, and heat carrier, are realized. Additional research should be carried out in case of poor results for some of the reliability characteristics (e.g., testing with and without material destruction, additional analysis of data related to the exploitation of the observed system or its analogues, etc.). Since the given procedure has iterative character, it is interrupted after confirming the initial hypothesis related to the agreement of the result of prognosis and actual exploitation data. The following is the process of developing a time plan for specific activities in order to achieve a higher level of reliability, using one of the standard methods (the method of network plan—Program Evaluation and Review Technique or PERT, the Gantt diagram or linear chart-chart, etc.), phase III of the process shown in Figure 1.11. Optimization according to the proposed algorithm of the modified method is based on the indirect method and internally in advance orientated specific range of variation of certain characteristics of reliability. This manner allows for certain simplification of a large part of the undetermined impacts and conditions of exploitation, that is, the very accuracy of this method is limited by the accuracy of the optimal results with properly given starting data, with the possibility of its continuous correction on the stage of exploitation of the object. Methods and programs for solving the reliability of certain parts of the technological scheme or the system as a whole, with adequate planning of the program, content, and duration of planned overhauls, are based mainly on statistical analysis and the use of analogue-based results. Models and methods for solving optimization tasks can be conditionally classified into several subgroups:

a. Increase of the influence of the basic technological scheme of the plant on the mono and double block structure without its detailing and analysis of losses in the structure of equipment, through analysis based on nominal power

b. Variant assessment of the scheme of plant through the application of the FMEA/FMECA analysis and the tree of failure of the most critical drive (obtained by the method of ranking), combined with the analysis of semi-Markov or Markov processes of failure according to the criterion of the condition without failures or without occurrence of breakdown or the readiness.

c. Variant assessments of a complex set of reliability indicators for the corresponding principled scheme and structural reservation of the plant based on previously performed assessment of reservation of a higher hierarchical level of the EES, with simplified or detailed analysis (depending on the predefined precision).

1.8 System of Maintenance with Entries and Exits Toward Environment

Reinforcement of connections between technical and technological complexity of thermal power plant (TPP) (in the form of a high initial price of the plant and equipment and more rational exploitation—production and economic factors, effect of breakdown on the reduction of production of electric and thermal energy, as well as technological gas on one side and ecological requirements on the other side) conditions a complex dependency of the maintenance function on a large number of factors. Reduction of possibilities in realizing the goal function set in the form of certain criteria is followed by the appropriate change of the status of system and its elements, being most often a consequence of wear, fatigue, corrosion, abrasion, pressure, heating, ageing, etc. The maintenance system of TPP is linked with the environment through certain entry parameters presented in Figure 1.12:

a. Characteristics of TPP as an overall system and its consisting elements and accessory equipment (kind of TPP, place and status of TPP within electro energetic system (EES), manner of connections through transmission network, place and role within the system of

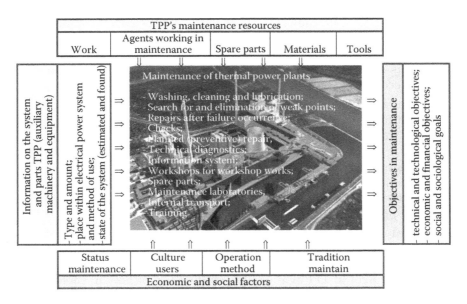

FIGURE 1.12
Presentation of the system of maintenance of TPP and ties with the environment. (Based on Milovanović [38].)

South East Europe, estimated status, manner of exploitation, status of the level of applied technical diagnostics, etc.)
b. Available maintenance resources (available personnel for operations, possibility of use of services of external specialized companies for individual segments of maintenance and individual actions in maintenance, operating tools for maintenance, spare parts and possibility of producing spare parts on its own, materials, tools, etc.)
c. Economic-social factors (place and status of maintenance in the organization of TPP, customer's culture, management of the system of TPP, tradition, etc.).

On the other hand, the system of maintenance of TPP with its organization structure, adopted strategy, and applied technologies of maintenance should satisfy planned objectives of maintenance (exit toward environment), defined within several groups:

a. Technical-technological objectives (maintenance and increase of working capacity up to the required level of efficiency of the system of TPP and its higher hierarchical system of EES, realization of planned basic and extended revitalized operating period of the plant and equipment of TPP, qualitative and quantitative improvement of the process of production of useful forms of energy, along with meeting general and special quality standards and service safety in the continuous supply of energy for the customer, increase of the production capability of TPP as a whole, continuous work on reconstruction, modernization and improvement of individual segments of TPP, with the goal of increasing technological, economic and ecological acceptability, etc.)
b. Economic-financial goals (reasonable use of the plant and equipment from the aspect of expenditure of spare parts, materials, tools and accessories, services of external specialized companies in individual maintenance jobs, then purchase of raw materials, available human potential, investment in maintenance in function of raising rentability, productivity and efficiency, decrease of costs with retaining the required level of maintenance in the value of TPP and its final product in the form of useful energy, etc.)
c. Social-sociological goals (development of the local community and broader region through the compensations linked with the work of TPP, reasonable use of primary energy resources through an optimal management of TPP within EES, rational use of human capacities, preservation and growth of psychological stability of employees, increase of motivation for work by raising the reliability of the system as a whole, ecological risk in order to protect environment, risk

at work in order to protect the personnel at TPP and population that lives and works in the closer and broader environment of TPP, etc.).

The concept of maintenance organization within TPP should provide a comprehensive approach to all maintenance activities, with the goal of realizing the defined goal function in the form of continued production of useful energy, along with a safe and sustainable work with regard to the personnel and environment. Concerning its dynamic character, the system of maintenance of TPP prepares and implements the maintenance function in the form of organized forms by individual groups of equipment and plants. Therefore, modern strategies for maintenance of TPP should in practice represent most optimally integrated group of adequate data bases (subsystems), like the system of equipment components (German Kraftwerk Kennzeichen System—KKS system), determined significance of individual systems or components for a reliable and safe plant of TPP as a whole (analysis of causes and consequences of failures, analysis of the failure tree, analysis of importance with regard to reliability of the system as a whole, methods for assessment of reliability and determination of significance, along with defining "bottlenecks" or "critical points," etc.), and system of potential mechanism of defects of individual components (CEN documents, e.g., CWA 15740:2008, etc.) and potential drive (operating) problems with TPP (statistics according to: API, ASME, OREDA, NERC, CEN, etc.). After defining the maintenance strategy (approach to maintenance), it is necessary to also define the *technological maintenance processes* (*maintenance technology*), with the goal of realizing set goals by an adequate maintenance strategy [45]. This implies elaboration of the maintenance technology itself, adoption of the known principles of performing recording of failures and repairs themselves, determination and diagnostics of different parameters that define the status of the system of TPP, as well as definition of appropriate repair technologies for the repair of damaged parts, lubrication, anticorrosive protection, etc. The decision on the type of maintenance of TPP is taken based on the criterion of company expenditures referring to maintenance and exploitation. In this manner it is possible to determine economically most acceptable activities of maintenance, which are necessary to be realized by applying an appropriate type of maintenance. In doing so, it is also not allowed to neglect the activities of maintenance technology linked with the control and diagnostics in maintenance, the repair technologies of maintenance themselves, as well as the activities connected with lubrication and anticorrosive protection of technical systems [46]. Among the procedures that are most frequently used for repairing broken or worn-out parts on TPP, those used are as follows: welding, built-up welding, metallization, electrolytic application, electromechanical processing, *Metalock* procedure of connecting broken parts, gluing technique, patented technologies for repair welding, application of material on the surface by the techniques of *Plasma Spraying* and *Flame Spraying*, etc.

1.9 Overview of Maintenance Activities in TPP

Important elements when determining the strategy for the maintenance and the scale and methodology of inspection of certain elements and the equipment in TPP are understanding and systematics of possible operative problems during exploitation (disturbance or complete loss of function), so as systematics of possible mechanisms of damage emergence (damaging or degradation of the material). In that light, according to the modified classification in the system OREDA or NERC, disturbances or deviations and problems related to structure materials in thermal power plant consist of several subgroups: fouling and deposits without fluid flow disturbances, fluid flow disturbances, like high or low fluid flow (HFF/LFF), no fluid flow (NFF), so as other fluid flow problems (OFFP), noise (NOIS), and vibrations (VIB), improper dimensioning and improper clearances, man-made disturbance, like deliberate disturbance, disturbance to insufficient training, etc., accidents, like fires, explosions and similar, improper start or stop—failed to start or stop (FTS), failed while running (FWR), external leakage (EXL) and overheating, thermal overheating (OHE), and other problems (OTH). Also, there are classifications regarding problems of installed material, Table 1.5. This kind of systematization enables grouping and marking cases of failures or possible damages, what as a consequence have easier statistical processing of the data on the level of individual parts or elements of the power plant (inside electro energetic system). Besides the history of operation (statistically processed data regarding past operation of the plant), for practical determining state of the system, it is necessary to create the plan of inspection by using convenient methods of technical diagnostics. Determining the scale of the damage and predicting its further progress together with following evaluation of risk and safety in the plant operation demands additional optimization of choosing the component and location of possible damage, type of the method for detection of the damage, but also the method for further estimation of scale and degree of the damage.

1.10 Unsolved Tasks and Directions of Further Developing Models for Forecasting Reliability of CTSs

The display of the developed reliability models mostly relies on the probability theory, with one part, to a greater or lesser extent, reflecting the experiences from practice or experimental researches. Other researches include empirical experience and experiment, but also have serious shortcomings despite the tendency to get closer to realistic conditions. For example, models that calculate a system reliability after preventative maintenance actions are

TABLE 1.5

Systematics of Problems and Damages by CEN CWA 15740:2008

Event, Issue	Identification and Type of Damage or Disturbances/ Deviations, Functional Problems	Subtypes/Specifics/Further Details/Examples
	Corrosion, Erosion and Environment Related Damage	
Material Damage Related Problems I	A. Volumetric loss of material on surface	General corrosion, oxidation, erosion, wear, extended thinning
		Localized pitting, crevice or galvanic corrosion
	B. Cracking on surface, mainly	Stress corrosion—chloride, caustic, etc., cracking
		Hydrogen induced damage, in blistering and HT hydrogen attack
		Corrosion fatigue
	C. Material weakening and/or embrittlement	Thermal degradation
		Carburization/decarburization, dealloying
		Embrittlement, incl. hardening, strain aging, temper embrittlement, liquid metal embrittlement, etc.
	Mechanical or Thermo-mechanical Loads Related	
Material Damage Related Problems II	A. Wear	Sliding wear
		Cavitational wear
	B. Strain, instability, dimensional changes, collapse	Overloading, creep
		Handling damage
	C. Microvoid formation	Creep
		Creep-fatigue
	D. Micro-cracking, cracking	Fatigue HCF and LCF, thermal fatigue, corrosion fatigue
		Thermal shock, creep, creep-fatigue
	E. Fracture	Overloading
		Brittle fracture
	Other Structural Damage Mechanisms	

Source: Data from Milovanović [39].

inadequate and difficult to be applied, while models based on the reliability assessment of complex repair systems often do not decompose the system, while considering the reliability of individual system components, and the system is viewed integrally giving a simplified and imprecise assessment of system reliability. On the other hand, the interaction between the failures of system components is not adequately modeled, and the existing models related to dependent failures consider mainly the one-way effects of failure (models with continuous interactions between components have not been sufficiently developed yet). There are no adequate models developed to evaluate the reliability of the system based on a small number of failures or in cases where they do not exist. This means, as with any model, that certain assumptions also create some limitations of the model, but the flexibility of the model can be a way that it can be applied to as many technical systems as possible, and that its limiting assumptions can be corrected in accordance with the needs of modeling the reliability of the concrete technical system. Previous researches have some yet unsolved tasks in front of them. For example, there is a very large inconsistency between theoretical research and applied models in practice. The majority of models have been developed by scientists in the form of theoretical, mathematical, statistical, informatics or other problematics which cannot be implemented and solve a practical problem in the industry. The task for the next period is to develop models that can be implemented to solve practical problems in the industry, as well as to create models that can be applied in the case of a small number of available system data, and especially a small number of failures (the reality for many technical systems and the fact that when there is no alternative). The accuracy of the reliability model is something that still provides great opportunities for improvement. This is especially important when it comes to reliability prediction, which is the basis of planning and decision-making on optimal maintenance actions according to the given criteria. The reliability models relating to the CTSs have not been developed yet as sufficiently precise, applicable, and they do not include important real factors (further work in these directions is necessary).

For engineering application it is necessary to know the methods of calculating, testing, and installing the reliability of complex systems, which include software (so-called software systems) in the development phase, on one side, but the knowledge of methods of increasing the reliability of such systems in the period of use is also required on the other side. The reliability of a complex system depends on the level of reliability of each of its components. There are mathematical relationships that show the dependency of the reliability of the complex system on the reliability of the components. The reliability of the system is installed in the development phase, and it is therefore called inherent reliability. If the system is well designed, tested in details, well maintained, with proper use, a high level of reliability in use is expected. However, the reliability of the system is significantly influenced by the environment. Reliability is measured and has a practical interpretation. The exact value of reliability is never known, but its numerical estimate of close real value can be obtained. Such an estimate can be obtained with stochastic methods based on data obtained by measuring at a particular set. The reliability estimate based on a certain number of data sets is expressed by the reliability level, which represents the mathematical probability and connects the estimated and actual (but unknown) value of reliability. The level of reliability is the probability that a certain value of reliability will be between the lower and upper bounds.

During the development of CTSs and artificial intelligence systems, it is often necessary to make decisions in terms of indeterminacy. Because of the nature of indeterminacy, it is basically impossible to predict the consequences of certain activities, technical solutions, failures, etc. with absolute reliability. The application of quantitative models is focused on the use of the notion of probability to describe the indeterminacy of a different nature. The so-called *Bayesian approach* for assessment of reliability and security proved to be very perspective. In Bayesian approach, the indeterminacy is seen as a probability that can be interpreted as a relative frequency, as a level of conviction or in some other way. To solve the problem of indeterminacy in this approach, it is necessary to provide a set of a priori probabilities that describe the basic set. The a priori probabilities can be determined by means of frequencies or statistical analyses. Such statistical analyses start in advance from the fact that the relevant data for describing the basic set are available. If such data are not available, then the a priori probabilities are given as subjective assessments by the experts. The result of the analysis represents a set of a posteriori probabilities. On the other hand, the theory of fuzzy sets is a convenient mathematical apparatus for the treatment of indeterminacy. A fuzzy set represents such a set (interval) of values with the corresponding function of belonging to the given in this interval. There is on one translation for the term fuzzy, and in many works, the fuzzy is translated as: blurry, unsharp, papery, fluffy, fibrous, imprecise set. However, most often this term is not translated, but used in its original form, that is, as a *fuzzy set*. In the case when the starting data for reliability analysis are not given point-to-point, the problem of estimating the degree of criticality can be successfully solved by setting the change interval with the corresponding belonging functions. This approach can be implemented in expert systems designed for reliability analysis and failure analysis.

1.11 Conclusion

The database of input data, in addition to basic data about plant, has to contain relevant data on the process of previous exploitation and maintenance. The most important feature of these data is the *truth*, because based on the subsequent statistical processing of the basic indicators a further strategy for the operation on this plant is determined. In order to monitor the functioning of the reference system and define its reliability indicators, power events in the EES over a specific time period (usually the previous period of operation) are monitored and recorded. On the basis of data from exploitation, the total duration of the units outside the plant is calculated, and on the level of the entire observed time period the following basic data: the number of failures and stoppages, the average time of stoppage, the causes of failures and stoppages, the unavailability of the components and the TPP as a whole, the total unsupplied electric energy. The reliability indicators of the observed TPP, as well as its individual units are calculated by statistical processing of the recorded data. A set of data developed by recording of driving events and the results of statistical processing of the same are referred to as "statistics of driving events." The statistics of driving events are also used for comparison with other analogue systems and the evaluation of performance of company that manages the TPP, as well as for studies of planning and probabilistic simulations of the operation of the system as a whole. Stoppage of units and components of TPP can be considered as random events joined by certain probability. Maintenance of machinery, equipment, and CTSs, from the aspect of the amount of necessary investments during their lifespan, is directly a function of defining and realizing the wanted efficiency (reliability, readiness, and suitability for maintenance), both at the level of their projecting and in the course of their exploitation itself. A well-chosen concept of maintenance, with a correct organization, programming, and realization of individual maintenance activities during exploitation, along with well-trained personnel and provided maintenance control, also affects improvement of economic results of the given organization or company. On the other hand, with the increase of complexity of technical systems there also occurs a problem of their optimal functionality, particularly if we know that such systems may often cause big economic losses or endanger security of a broader macro region and people serving them. Each CTS carries within it a big potential danger of possible occurrence of failures and break-downs dangerous for a broader environment. Reliability of complex systems, designed such as to successfully perform the function, determines lasting of the time interval in which the system will function without a failure. The research directed toward the increase of the reliability level and management of reliability during the lifespan of the object aims at defining the system of protection measures and their optimization from the aspect of simultaneous provision of exploitation efficiency and realization of complex regulations linked with the environment protection and safety of both micro and macro region. Special tasks of maintenance technology are provision of the process of maintenance optimization and advancement of principles for achieving higher quality, reliability, and efficiency of the CTS and its own production. The decision on the type and activities of maintenance may be taken also on the basis of expenditures of the company referring to maintenance and exploitation and chosen methods for maintenance. The order of development steps, which need to be implemented, affects both efficiency and effectiveness of the maintenance system. The formulation of tasks related to the optimization of the reliability of the thermal power plant system can in general be defined as the minimization of the loss on the construction and application of a serial unified power plant, which consists of losses associated with the installation itself and losses due to its connection to the ESS, depending on the reliability indicators and the possible manners for their provision for the given takas at each stage of the life cycle, as well as the taken system parameters and the known minimum necessary functional structure of the plant. For these reasons, all limitations of operation of the thermal power plant within the electric power system are also valid. Solving this problem is, taking all the given limitations and the overall analysis of the interconnection and overlapping of certain expressions, pretty complex from the aspect of the totality of the equation system, and for these reasons considerable simplification is done for the level of assessment of reliability indicators. Sometimes, in the process of optimization, the costs related to the hierarchical connection of the thermal power plant with the environment and the measures for implementing its supplementary protection, as well as the occurrence of possible restrictions, are included. The next step is the grouping of the said costs per objects, life cycle stages, and purpose of resources. This system should be supplemented with costs related to providing reliability for each of the phases of the life cycle of the thermal power plant system.

References

1. Papić, Lj., Milovanović, Z., Systems Maintainability and Reliability, The DQM monography library "Quality and Reliability in Practice", Book 3, The Researching Center for Quality and Reliability Management, Prijevor, 2007.
2. Milovanović, Z., Optimization of thermo power plants liability. Faculty of Mechanics, University in Banja Luka, Banja Luka, 2003.
3. Hoyland, A., Rausand, M., *System Reliability Theory: Models and Statistical Methods*. New York: John Wiley & Sons, Inc., 1994.
4. Sun, Y., Reliability prediction of complex repairable systems: An engineering approach. Thesis submitted in total fulfilment of requirements of the degree of Doctor of Philosophy, Faculty of Built Environment and Engineering, University of Technology, Queensland, 2006.
5. Dev N., Samsheb, Kachhwaha S.S., Attri R., Development of reliability index for combined cycle power plant using graph theoretic approach, *Ain Shams Engineering Journal, Ain Shams University*, 2014, 5, pp. 193–203.
6. Naess, A., Leira, B.J., Batsevych, O., System reliability analysis by enhanced Monte Carlo simulation, *Structural Safety*, 2009, 31, pp. 349–355.
7. Weber, P., Jouffe, L., Complex system reliability modelling with Dynamic Object Oriented Bayesian Networks (DOOBN), *Reliability Engineering and System Safety*, 2006, 91, pp. 149–162.
8. Moazzami, M., Hemmati, R., Haghighatdar Fesharaki, F., Rafiee Rad, S., Reliability evaluation for different power plant busbar layouts by using sequential Monte Carlo simulation, *Electrical Power and Energy Systems*, 2013, 53, pp. 987–993.

9. Petrović, G., Marinković Z., Marinković, D., Optimal preventive maintenance model of complex degraded systems: A real life case study, *Journal of Scientific & Industrial Research*, 2011, 70, pp. 412–420.
10. Milošević, A., Reliability ensuring models of complex facilities in thermal power plants. Thesis submitted in total fulfilment of requirements of the degree of Doctor of Technical Science (Industrial Engineering), University in Novi Sad, Technical Faculty "Mihajlo Pupin", Zrenjanin, Serbia, 2012.
11. Milovanović, Z., Modified method for reliability evaluation of condensation thermal electric power plant. Doctoral thesis, Faculty of Mechanical Engineering Banja Luka, Banja Luka, 2000, pp. 180–229.
12. Kalos, M.H., Whitlock, P.A., *Monte Carlo Methods*. New York: John Wiley & Sons, 1986.
13. Finkelstein, M.S., A point-process stochastic model for software reliability, *Reliability Engineering & System Safety*, 1999, 63(1), pp. 67–71.
14. Fiems, D., Steyaert, B., Bruneel, H., Analysis of a discrete-time GI-G-1 queuing model subjected to burst interruptions, *Computers & Operations Research*, 2003, 30(1), pp. 139–153.
15. Cox, D.R., Oakes, D., *Analysis of Survival Data*. London: Chapman & Hall, 1984, pp. 91–113.
16. Ebeling, C.E., *An Introduction to Reliability and Maintainability Engineering*. New York: The McGraw-Hill Company, 1997, pp. 124–128.
17. Mazzuchi, T.A., Soyer, R.A., Bayesian perspective on some replacement strategies, *Reliability Engineering & System Safety*, 1996, 51(3), pp. 295–303.
18. Sheu, S.H., Yeh, R.H., Lin, Y.B., Yuang, M.G., A Bayesian approach to an adaptive preventive maintenance model. *Reliability Engineering & System Safety*, 2001, 71(1), pp. 33–44.
19. Percy, D.F., Kobbacy, K.A.H., Fawzi, B.B., Setting preventive maintenance schedules when data are sparse. *International Journal of Production Economics*, 1997, 51(3), pp. 223–234.
20. Apeland, S., Scarf, P.A., A fully subjective approach to modeling inspection maintenance. *European Journal of Operational Research*, 2003, 148(2), pp. 410–425.
21. Liu, Z., Liu, Y., Cai, B., Zhang, D., Zheng, C., Dynamic Bayesian network modeling of reliability of subsea blowout preventer stack in presence of common cause failures, *Journal of Loss Prevention in the Process Industries*, 2015, 38, pp. 58–66.
22. Tian, Z., Liao, H., Condition based maintenance optimization for multi-component systems using proportional hazards model, *Reliability Engineering & System Safety*, 2011, 96(5), pp. 581–589.
23. Belitser, E., Serra, P., Zanten, H., Rate-optimal Bayesian intensity smoothing for inhomogeneous Poisson processes, *Journal of Statistical Planning and Inference*, 2015, 166, pp. 24–35.
24. Lee, M., Sohn K., Inferring the route-use patterns of metro passengers based only on travel-time data within a Bayesian framework using a reversible-jump Markov chain Monte Carlo (MCMC) simulation, *Transportation Research Part B: Methodological*, 2015, 81(1), pp. 1–17.
25. Liu, Y., Li, C., Complex-valued Bayesian parameter estimation via Markov chain Monte Carlo, *Information Sciences*, 2016, 326(1), pp. 334–349.
26. He, Z., Gong, W., Xie, W., Zhang, J., Zhang, G., Hong, Z., NVH and reliability analyses of the engine with different interaction models between the crankshaft and bearing, *Applied Acoustics*, 2016, 101(1), pp. 185–200.

27. Yi, C., Bao, Y., Jiang, Y., Xue, Y., Modeling cascading failures with the crisis of trust in social networks, *Physica A: Statistical Mechanics and its Applications*, 2015, 436(15), pp. 256–271.
28. Henneaux, P., Probability of failure of overloaded lines in cascading failures, *International Journal of Electrical Power & Energy Systems*, 2015, 73, pp. 141–148.
29. Duan, D., Ling, X., Wu, X., OuYang, D., Zhong, B., Critical thresholds for scale-free networks against cascading failures, *Physica A: Statistical Mechanics and its Applications*, 2014, 416, pp. 252–258.
30. Cupac, V., Lizier, J.T., Prokopenko, M., Comparing dynamics of cascading failures between networkcentric and power flow models, *International Journal of Electrical Power & Energy Systems*, 2013, 49, pp. 369–379.
31. Wu, X., Wu, X., Extended object-oriented Petri net model for mission reliability simulation of repairable PMS with common cause failures, *Reliability Engineering & System Safety*, 2015, 136, pp. 109–119.
32. Greig, G.L., Second moment reliability analysis of redundant systems with dependent failures, *Reliability Engineering & System Safety*, 1993, 41(1), pp. 57–70.
33. Mosleh, A., Common cause failures: An analysis methodology and examples, *Reliability Engineering & System Safety*, 1991, 34(3), pp. 249–292.
34. Sun, Y., Ma, L., Mathew, J., Zhang, S., An analytical model for interactive failures, *Reliability Engineering & System Safety*, 2006, 91(5), pp. 495–504.
35. "AMSAA Design for Reliability Handbook", Technical Report No. TR-2011–24 August 2011. US Army Materiel Systems Analysis Activity Aberdeen Proving Ground, Maryland 21005-5071 Approved, Page 10.
36. Miličić, D., Milovanović, Z., Library monographs: Energy—Generating machines steam turbines. Faculty of Mechanics, University in Banja Luka, Banja Luka, 2010.
37. Milovanović Z., Library monographs: Power and process plants, Volume 1: Thermal power plants—Theoretical basis. Faculty of Mechanics, University in Banja Luka, Banja Luka, 2011.
38. Milovanović Z., Library monographs: Power and process plants, Volume 2: Thermal power plants—Technological systems, design and construction, operation and maintenance. Faculty of Mechanics, University in Banja Luka, Banja Luka, 2011.
39. Milovanović Z., Library monographs: Energy—Generating machines thermodynamic and flow dynamics basics of thermal turbo machiners. Faculty of Mechanics, University in Banja Luka, Banja Luka, 2010.
40. Каплун С.М., Оптимизация надежности энерго установок, Отв. ред. Г. Б. Левенталь, Акад. наук СССР, Сиб. отделение. Сиб. энерг. инцтитут, Наука, Новосибирск, 1982, pp. 200–272.
41. Pavlović N., Time energy indicators of reliability of thermal units in the electric power industry system of Yugoslavia and the comparison with the region's UNIPEDE. Faculty of Mechanical Engineering, Beograd, 1986.
42. Розанов М.Н., "Надежность электро-энергетических систем", Справочник, Том 2, Энергоатомиздат, Москва, 2000.
43. Milovanović Z., Knežević D., Milašinović A., Dumonjić-Milovanović S., Ostojić D., Modified method for reliability evaluation of condensing thermal power plant, *Journal of Safety Engineering*, 2012, 1(4), pp. 57–67.

44. Milovanović Z., Dumonjić-Milovanović S., Reliability assessment of condensing thermal power plants, technique—Mechanics, *Union of Engineers and Technicians of Serbia*, 2015, 1(64), pp. 86–94
45. Milovanović Z., Dumonjić-Milovanović S., Branković D., Models for achieving cost-effectiveness and sustainability during exploitation and maintaining of thermal energetic facility, *2nd Maintenance Forum on Maintenance and Asset Management, Job of Maintenance Community, Conference Proceedings*, Montenegro, 2017, pp. 204–219.
46. Milovanović Z., Milašinović A., Knežević D., Škundrić J., Dumonjić-Milovanović S., Evaluation and monitoring of condition of turbo generator on the example of thermal power plant Ugljevik 1×300 MW, *American Journal of Mechanical and Industrial Engineering*, 2016, 1(3), pp. 50–57.

2

Non-exponential Distributions in Reliability Modeling of PMS: Approximation and Simulation Approaches

Xiang-Yu Li, Jun Hu, Hong-Zhong Huang, and Yan-Feng Li
University of Electronic Science and Technology of China

CONTENTS

2.1 Introduction: PMS and Non-exponential Distributions 48
2.2 Approximation and Simulation Methods for Dynamic
    Non-exponential Systems 49
    2.2.1 Approximation Approach 50
    2.2.2 Simulation Approach 52
        2.2.2.1 Basic Conceptions of Monte Carlo Simulation............ 52
        2.2.2.2 Simulation Procedure 53
    2.2.3 Numerical Example 54
        2.2.3.1 Approximation Approach.. 54
        2.2.3.2 Simulation Approach and the Results 55
        2.2.3.3 Different Non-exponential Distributions Analysis 56
2.3 Non-exponential PMS Analysis 60
    2.3.1 PMS with Partially Repairable Non-exponential
    Components 60
        2.3.1.1 The AOCS System with Partially Repairable
            Components 60
        2.3.1.2 The Modular Approach... 62
        2.3.1.3 Module and System Reliability 64
    2.3.2 PMS Analysis Considering Random Shocks Effect................ 69
        2.3.2.1 MRGP and Multistate Random Shocks Model 69
        2.3.2.2 Simulation Procedure 71
        2.3.2.3 Case Study... 73
2.4 Conclusion 77
References... 78

2.1 Introduction: PMS and Non-exponential Distributions

The phased mission systems (PMSs) are systems subject to consecutive, multiple, nonoverlapping mission time durations (phases) and they need to complete different missions. Such a system is usually seen in aerospace, nuclear power, and many other applications [1]. One classic example of the PMSs is the manned spacecraft which involves takeoff, orbital transfer, orbit operation, and back to earth phases. In each phase, the system needs to accomplish different missions, which means that the system structure and failure criteria differ from phase to phase [2–4]. Furthermore, the system may be subject to different stresses in different phases due to different environments, for example, the endo-atmosphere in the takeoff phase and the outer space in the orbital operation phase of the spacecraft. Therefore, a distinct model for each phase is necessary for the modeling and assessment of an accurate system. But with the distinct models, the dependences across the phases pose a great challenge to the reliability modeling and assessment of the whole system; for example, in a non-repairable PMS a component that fails in the former phases will also stay in the failure state in the latter phases, making system modeling and assessment more difficult. Furthermore, many practical systems are subject to dynamic behaviors, such as cold standby or functional standby [5]. Therefore, the dependences across the phases and dynamic behaviors pose great challenges to the existing system modeling and assessment methods.

To deal with the phase dependence, many works, like the reliability analysis of the phased satellite and spacecraft, have been done on the reliability analysis of PMSs in the past few decades. The PMS can be static or dynamic [6]. If the failure of the mission in any phase is only dependent on the combinations of component failure events, the PMS is static. In a static PMS, the system structure of each phase can be represented by a static fault tree (FT) model; that is, all the logic gates are static gates (OR, AND, or k-out-of-n). If the failure order of components affects the system state, the PMS is said to be dynamic, such as cold standby. The FT model of the dynamic phase contains at least one dynamic gate (priority, standby, or functional dependent) [7,8].

According to the system behaviors, the existing works can be divided into two major categories:

1. Combinatorial methods (BDD-Binary decision diagram, MDD-Multi-valued decision diagram, or MMDD-Multi-state multi-valued decision diagram based models) [1–6,8–10]: The BDD method was proposed by Zang and Trivedi [2] to assess the system reliability of the PMSs. Xing applied the BDD method in reliability analysis of the generalized PMS [3]. He also used the BDD method to assess the system reliability of the PMS considering common cause failure (CCF) and imperfect coverage [1] or considering both internal and external CCF [4]. Tang and Dugan assessed the system reliability of

the PMS considering multimode failures [9]. Besides the BDD model, the MDD model is used in the PMS reliability analysis, especially in the PMS considering multi-failure modes. Mo [10] showed that the MDD model is more efficient than the BDD model in the PMS with multi-failure modes. The combinatorial method can provide highly efficient system reliability, especially in large-scale systems. However, this method can only be used if the system is combinatorial (i.e., the bottom events are independent of each other).

2. State-space-oriented model (Markov chain-based or Petri-nets-based approaches) [5,11,12]: This kind of model can deal with the dynamic behaviors within phases like functional dependent or cold spare (CSP) but they suffer from the state explosion problem.

To avoid the limitations of these two methods, a modularization method was proposed [13]. Also, Zhu and Lombardi proposed a stochastic computation method to evaluate the reliability of the PMS more efficiently [14].

Most of these existing research works assumed that the lifetimes of the system components follow the exponential distributions, which are commonly used in reliability modeling due to its memoryless property, especially in the continuous-time Markov chains (CTMC) [15]. But in practice, most of the components or subsystems follow the non-exponential distributions, for example, the Weibull distribution [16] or the log-normal distribution [17] which is not available in the CTMC. In the PMSs such as spacecraft, many components or subsystems are mechanical or electromechanical whose lifetimes more likely follow the non-exponential distributions. To deal with the non-exponential distributions in the system with dynamic behaviors, the non-Markovian models including the semi-Markov process (SMP) or the Markov regenerative process (MRGP) [17–21] are required.

This chapter will provide a state-of-the-art review of the evaluation methods of the non-exponential dynamic systems and the modular approach for the PMS. In the traditional PMS, lifetimes of all the components are assumed to follow the exponential distributions, which is not reflected in reality. In this chapter, the modeling and assessment methods for the PMS cover the non-exponential distributions as well as the approximation approach and simulation approach. Methods to consider partially repairable and random shocks in the reliability analysis of the PMS will also be discussed in this chapter.

2.2 Approximation and Simulation Methods for Dynamic Non-exponential Systems

Any traditional combinatorial method like the BDD method or the MMDD method is not available for the analysis of a dynamic PMS, so the state-space models are used to construct the system model for each phase, for example, the CTMC and Petri-net models. The CTMC can deal with the state-space models with the transition time exponentially distributed. But in practice, the lifetimes of most components do not follow the exponential distributions. Therefore, the non-Markovian dependability modeling methods including the SMP and MRGP are necessary. But due to the complexity of the evaluation process, these models are not widely used. In this section, two evaluation approaches (the approximation method and the simulation method) for the non-Markovian dependability modeling methods are shown in detail.

2.2.1 Approximation Approach

A continuous-time stochastic process is called an SMP if the embedded chain is a Markov chain and the transition time between two states follows arbitrary distributions [22]. The SMP is a generalization of the classic Markov chains as it accommodates arbitrary sojourn time distributions. Generally, the SMP does not have the Markov property, except for transition time points. These time points are the Markov regeneration epochs, and the SMP only changes the states at these epochs. That is why it is called an SMP [22–24]. Figure 2.1 shows an SMP model without repair.

$F_{i,j}(t)$ is the cumulative distribution function (CDF) of the transition time between system states i and j, and $F_{i,j}(t)$ can be any distribution in an SMP with its parameters $\lambda_{i,j} = \left[\lambda_{i,j}^1, \ldots, \lambda_{i,j}^n\right]$.

To define the transition behaviors of an SMP, let the initial system state probability vector be $P(t)$ at time $t = 0$ and the kernel matrix $Q(t)$ in which element $Q_{i,j}(t)$ denotes the probability that the SMP transitions from state i to state j during the time interval $[0,t]$ with one transition step. The kernel matrix $Q(t)$ can be obtained by the CDF of the sojourn time between states and the competition behaviors among transitions.

The main purpose of using the state-space model for reliability assessment is to evaluate the system state probabilities at any time t. Let $\theta(t)$ represent the transition probability matrix in which $\theta_{i,j}(t), i,j = \{1,2,\ldots,K\}$ represents the probability that the process starts from state i to state j in the time interval $[0,t]$. According to [23], the state probabilities

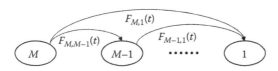

FIGURE 2.1
The transition diagram for the semi-Markov process.

$\theta_{i,j}(t), i,j = \{1,2,\ldots,K\}$ can be derived by solving the integrals given below [23]:

$$\theta_{i,j}(t) = \sigma_{i,j}\left(1 - F_i(t)\right) + \sum_{k=1}^{K} \int_0^t q_{i,k}(\tau)\theta_{k,j}(t-\tau)d\tau \quad (2.1)$$

where $q_{i,k}(t) = \dfrac{dQ_{i,k}(t)}{dt}$, $F_i(t) = \sum_{j=1}^{K} Q_{i,j}(t)$, $\sigma_{i,j} = \begin{cases} 1, & \text{if } i = j \\ 0, & \text{if } i \neq j \end{cases}$.

It can be observed that the first part of Eq. 2.1 is the probability that the system stays in state i at time interval $[0,t]$ and the second part of the equation represents the probability that the system transitions from state i to state j at time interval $[0,t]$.

By integrating the calculated $\theta_{i,j}(t)$ and the given initial system state probabilities, the system state probabilities at time t can be evaluated. With the system state probabilities at any time t, all the system reliability indices can be evaluated easily.

Although the SMP can deal with the non-exponential distributions, it is not widely used in reliability engineering. One important reason is that the integrals in Eq. 2.1 cannot be solved analytically under non-exponential distributions (e.g., Weibull distribution). To solve the complex integrals in Eq. 2.1, an approximation method based on the Trapezoidal Integral Law [25–27] is applied as follows:

$$\int_0^t q_{i,k}(\tau)\theta_{k,j}(t-\tau)d\tau \approx \frac{1}{2}\sum_{i=1}^{n-2}\left[q_{i,k}(\tau_i)\theta_{k,j}(t-\tau_t)\right]$$
$$\times \left[q_{i,k}(\tau_{i+1})\theta_{k,j}(t-\tau_{i+1})\right] \quad (2.2)$$

where $[0,t]$ is divided into n equal segments, so that the length of each segment is $\delta = t/n$. The higher the value of n, the more accurate the results.

With Eq. 2.2, Eq. 2.1 can be expressed as

$$\theta_{i,j}(t) \approx \sigma_{i,j}\left(1 - F_i(t)\right) + \frac{1}{2}\sum_{i=1}^{n-2}\left[q_{i,k}(\tau_i)\theta_{k,j}(t-\tau_t)\right]$$
$$\times \left[q_{i,k}(\tau_{i+1})\theta_{k,j}(t-\tau_{i+1})\right]. \quad (2.3)$$

With Eq. 2.3, all the system state probabilities can be evaluated at any time t recursively.

2.2.2 Simulation Approach

2.2.2.1 Basic Conceptions of Monte Carlo Simulation

The Monte Carlo (MC) simulation method for reliability assessment is based on the repeated sampling of realizations of system state configurations and the computation of the frequency of failure events [28,29]. The simulation method can also be used to test the correctness of the approximation solutions.

The key theoretical construct upon which MC simulation is based is the transition probability density function, which is defined as [30].

$f_{i,j}(\tau|t,\lambda)d\tau$ = the probability that, given the system arrives at the system state i at time t and transitions to system state j, that transition will occur in time interval $[t+\tau, t+\tau+d\tau]$.

λ is the vector of the parameters of the time in which the system transitions to state i at time t. According to the definition and previous introduction, $f_{i,j}(\tau|t,\lambda)d\tau$ can be expressed as

$$f_{i,j}(\tau|t,\lambda)d\tau = \Pr_i(\tau|t,\lambda) \cdot \theta_{i,j}(\tau,\lambda)d\tau \qquad (2.4)$$

where $\Pr_i(\tau|t,\lambda)$ is the probability that the system arrives in state i at time t with factors λ. Also no transition occurs in time interval $[t,t+\tau]$. Therefore, it satisfies

$$d\Pr_i(\tau|t,\lambda)/\Pr_i(\tau|t,\lambda) = -\theta_i(\tau,\lambda)d\tau \qquad (2.5)$$

where $\theta_i(\tau,\lambda)d\tau$ is the conditional probability that the system in state i at time t after arrived at state i at time $t-\tau$ and depart to other states at time interval $[t,t+\tau]$. Therefore, it can be calculated as

$$\theta_i(\tau,\lambda)d\tau = \sum_{j=1}^{M}\theta_{i,j}(\tau,\lambda)d\tau. \qquad (2.6)$$

with the obvious initial condition $\Pr_i(0|t,\lambda)=1$, $\Pr_i(\tau|t,\lambda)$ can be obtained as

$$\Pr_i(\tau|t,\lambda) = \exp\left[-\int_0^\tau \theta_i(\tau,\lambda)d\tau\right]. \qquad (2.7)$$

By integrating Eq. 2.7 into Eq. 2.4, we can obtain

$$f_{i,j}(\tau|t,\lambda)d\tau = \exp\left[-\int_0^\tau \theta_i(\tau,\lambda)d\tau\right]\cdot\theta_{i,j}(\tau,\lambda)$$

$$= \left(\theta_i(\tau,\lambda)\exp\left[-\int_0^\tau \theta_i(\tau,\lambda)d\tau\right]\right)\cdot\left(\frac{\theta_{i,j}(\tau,\lambda)}{\theta_i(\tau,\lambda)}\right)$$

$$= \psi_i(\tau|\lambda)\pi_{i,j}(\tau|\lambda). \qquad (2.8)$$

Reliability Modeling of PMS

$\psi_i(\tau|\lambda) = \theta_i(\tau,\lambda)\exp\left[-\int_0^\tau \theta_i(\tau,\lambda)d\tau\right]$ is the probability that the system state stays in state i at time t and leaves state i in time interval $[t, t+\tau]$. $\pi_{i,j}(\tau|\lambda) = \theta_{i,j}(\tau,\lambda)/\theta_i(\tau,\lambda)$ determines that the transition arrives state j after leaving state i. In the MC simulation procedure, $\psi_i(\tau|\lambda)$ is used to generate the holding time for system state i and $\pi_{i,j}(\tau|\lambda)$ is used to determine the arrival state. This procedure is repeated until the system reaches the system failure state.

2.2.2.2 Simulation Procedure

To generate the holding time τ among states, a two-stage simulation procedure is applied, as shown in Figure 2.2.

The probability of each system state changes over time $P_i(t)$ which can be calculated by statistically analyzing the time during which the system remains in a particular state.

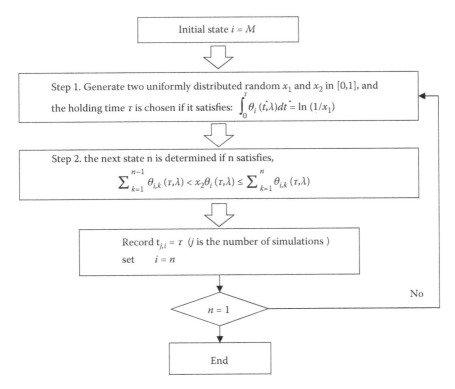

FIGURE 2.2
The Monte Carlo simulation procedure for the SMP.

2.2.3 Numerical Example

In this section, a system with one working component (A) and one cold standby component (B) is used to show the approximation method, as shown in Figure 2.3. In the system, only the component A can be repaired because of the limited maintenance resource due to weight restriction, which is commonly seen in aerospace equipment. Assume that component B should work immediately after the failure of component A and the switchover time is negligible.

In Figure 2.3, w, s, and f, respectively, denote a component staying at working, standby, and failure states and only system state 4 is the failure state. The parameters are shown in Table 2.1. α is the shape parameter and β is the scale parameter. The initial state probabilities are $\boldsymbol{P}(0) = (p_0(t) = 1, p_1(t) = p_2(t) = p_3(t) = 0)$.

2.2.3.1 Approximation Approach

From Figure 2.3, the $\boldsymbol{Q}$ matrix and the $\boldsymbol{\theta}$ matrix of the SMP for this example system can be represented as

$$\boldsymbol{Q}(t) = \begin{Bmatrix} 0 & Q_{1,2}(t) & 0 & 0 \\ 0 & 0 & Q_{2,3}(t) & Q_{2,4}(t) \\ 0 & 0 & 0 & Q_{3,4}(t) \\ 0 & 0 & 0 & 0 \end{Bmatrix} \quad (2.9)$$

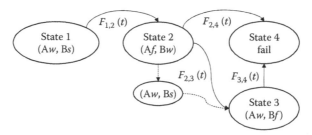

FIGURE 2.3
The State transition diagram for a cold standby system with partially repair.

TABLE 2.1
Parameters for the Example System in Section 2.2

	$F_A(t)$	$F_B(t)$	$G_A(t)$
α	2	1.5	1.5
β	10	10	20

Reliability Modeling of PMS

$$\boldsymbol{\theta}(t) = \begin{Bmatrix} \theta_{1,1}(t) & \theta_{1,2}(t) & \theta_{1,3}(t) & \theta_{1,4}(t) \\ 0 & \theta_{2,2}(t) & \theta_{2,3}(t) & \theta_{2,4}(t) \\ 0 & 0 & \theta_{3,3}(t) & \theta_{3,4}(t) \\ 0 & 0 & 0 & \theta_{4,4}(t) \end{Bmatrix}. \quad (2.10)$$

According to the competing behaviors and the CDFs of transition time, the $\boldsymbol{Q}$ matrix can be further ascertained as

$$\boldsymbol{Q}(t) = \begin{Bmatrix} 0 & F_A(t) & 0 & 0 \\ 0 & 0 & \int_0^t G_A(u)dF_B(u) & \int_0^t (1-G_A(u))dF_B(u) \\ 0 & 0 & 0 & F_A(t) \\ 0 & 0 & 0 & 0 \end{Bmatrix}. \quad (2.11)$$

By applying the Markov renewal equation, Eq. 2.1, and the approximation method in Section 2.2.1, all the elements in $\boldsymbol{\theta}(t)$ can be evaluated recursively. With the $\boldsymbol{\theta}(t)$ and initial state probability, the system state probability vector, $P(t)$, can be evaluated as

$$\boldsymbol{P}(t) = \boldsymbol{P}(0) \cdot \boldsymbol{\theta}(t). \quad (2.12)$$

2.2.3.2 Simulation Approach and the Results

As mentioned in the previous section, the results evaluated by the approximation method can be validated by the simulation method. The system state probabilities are evaluated by the MC simulation approach and the results are compared using the approximation method. First, $N_1(N_1 = 500)$ examples are generated by the MC simulation procedure given in Section 2.2.2. As is well known, with the increase of the amount of simulation data, the simulation result gets closer to the true value. Therefore, the amount of simulated data is increased to $N_2 = 5 \times 10^3$, $N_3 = 5 \times 10^4$, $N_4 = 5 \times 10^5$ and $N_5 = 5 \times 10^6$ separately. The max error and the mean error between the approximation method and the simulation method are shown in Table 2.2. The system reliability comparison between the approximation method and the MC simulation method with different amounts of data is shown in Figure 2.4. The max error and the mean error of the system reliability under different data amounts are shown in Table 2.2 as well.

From Table 2.2, we can see that the max error and the mean error between these two methods get smaller with an increase in the simulation amount.

TABLE 2.2

The Errors between the Simulation Method and the Approximation Method

Simulation Data Amount	5×10^2	5×10^3	5×10^4	5×10^5	5×10^6
Max error	4.96×10^{-2}	1.98×10^{-2}	4.37×10^{-3}	2.75×10^{-3}	4.49×10^{-4}
Mean error	1.10×10^{-2}	4.27×10^{-3}	1.04×10^{-3}	8.16×10^{-4}	2.66×10^{-4}
Calculation time of the simulation method(s)	0.69	5.37	60.59	773.44	6062.06

The results also show that the approximation method can provide a relatively accurate approximate solution. On the other hand, from the errors and calculation time shown in Table 2.2, we can see that finding an approximation solution by the MCS method will be more time-consuming than the approximation method. But the simulation method can deal with more situations than the approximation method.

2.2.3.3 Different Non-exponential Distributions Analysis

There are many non-exponential distributions that are used to describe the failure behaviors of components and systems in reliability research, for example, the Weibull distributions introduced above, the lognormal distribution or the uniform distribution, etc. In the previous sections, only the use of Weibull distribution has been shown as an example. The use of lognormal distribution and uniform distribution as lifetime distributions of components to show the generality of the approximation method is also shown in this section. The CDF and PDF of the uniform distribution and the lognormal distribution are

$$f_{unif}(t) = 1/(b-a), \quad a \le t \le b \tag{2.13}$$

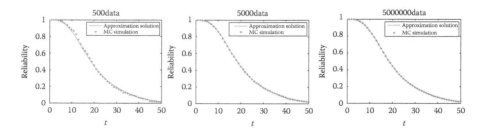

FIGURE 2.4
The comparison between the MCS and the approximation method with different data amount.

$$F_{unif}(t) = \begin{cases} 0, & t \leq a \\ (t-a)/(b-a), & a \leq t \leq b \\ 1, & t \geq b \end{cases} \quad (2.14)$$

$$f_{logn}(t) = \frac{1}{t}\frac{1}{\sigma\sqrt{2\pi}}\exp\left(-\frac{(\ln t - u)^2}{2\sigma^2}\right) \quad (2.15)$$

$$F_{logn}(t) = \Phi\left(\frac{\ln t - u}{\sigma}\right). \quad (2.16)$$

The a, b in Eqs. 2.13 and 2.14 and μ, σ in Eqs. 2.15 and 2.16 are the parameters for the uniform distribution and lognormal distribution, respectively. The curve of the uniform distribution with parameters $a = 5, b = 40$ and the lognormal distribution with parameters $\mu = 1.5, \sigma = 0.4$ are shown in Figures 2.5 and 2.6, respectively.

With the same example used above and the parameters shown in Tables 2.3 and 2.4 for uniform distributions and lognormal distributions of

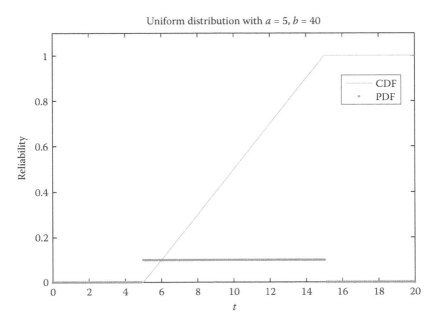

FIGURE 2.5
The PDF and CDF curve of the uniform distribution with $a = 5, b = 40$.

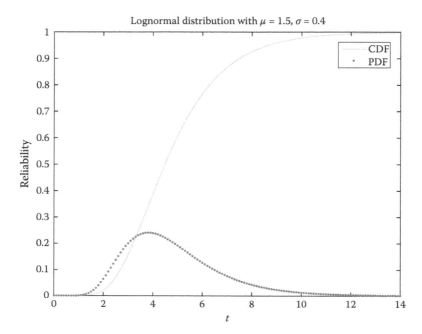

FIGURE 2.6
The PDF and CDF curve of the lognormal distribution with $\mu = 1.5$, $\sigma = 0.4$.

TABLE 2.3

Parameters for the Example System in Section 2.2 with Uniform Distribution

	$F_A(t)$	$F_B(t)$	$G_A(t)$
a	2	1.5	1.5
b	10	10	20

TABLE 2.4

Parameters for the Example System in Section 2.2 with Uniform Distribution

	$F_A(t)$	$F_B(t)$	$G_A(t)$
u	1.5	1	2.0
σ	0.7	0.8	0.9

the state transition time, the system reliability distribution type can be evaluated. The comparisons of the results of the MC simulation approach and the approximation approach under the uniform distribution and lognormal distribution are shown in Figures 2.5 and 2.6, respectively. And the comparisons of the approximation approach (the length of each segment $\delta = 0.1$)

Reliability Modeling of PMS

with simulation approach (the simulation amount $N_{max} = 2 \times 10^5$) for the example system in Section 2.2 with different distributions are shown in Figures 2.7 and 2.8, respectively.

The max error and the mean error of reliability of the same system with different lifetime distributions between the simulation approach (simulation

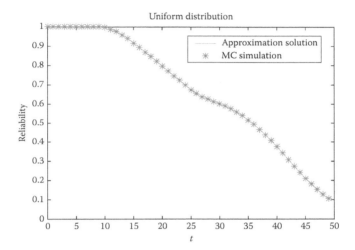

FIGURE 2.7
The comparison between the MCS and the approximation method with the uniform distributions of the example in Section 2.2.

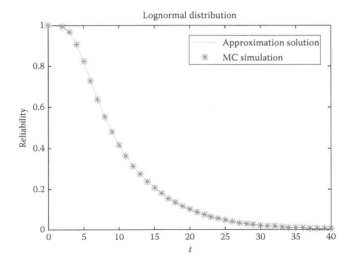

FIGURE 2.8
The comparison between the MCS and the approximation method with the lognormal distributions of the example in Section 2.2.

amount $N_{max} = 2 \times 10^5$) and the approximation approach (the length of each segment $\delta = 0.01$) and the computation time are shown in Table 2.5.

From Table 2.5, we can see that the approximation method can deal with not only the non-exponential distributions with Weibull distribution but also other non-exponential distributions. Furthermore, compared with the simulation approach, the computation efficiency of the approximation approach is much better. But the simulation method has more generality in analyzing components or systems, for example, the state of a component or a system can be transitioned to any other state with non-exponential distributions.

2.3 Non-exponential PMS Analysis

PMSs have wide applications in engineering practices, especially in the aerospace industry such as man-made satellites and manned spacecrafts. In these complicated aerospace systems, most of the components are mechanical components or mechatronics whose lifetimes follow non-exponential distributions like the Weibull distribution [31–33]. In this section, the dynamic behaviors are analyzed in the PMSs with non-exponential components, and the SMP and MRGP as well as the approximation approach and simulation approach are used. The PMSs with partially repairable components and the PMSs subject to random shocks are discussed in detail.

2.3.1 PMS with Partially Repairable Non-exponential Components

2.3.1.1 The AOCS System with Partially Repairable Components

As is well known, many real-world systems, particularly aerospace equipment like man-made satellites, are designed with cold-standby redundancy for achieving fault tolerance and high reliability [34,35]. On the other hand, due to the weight restriction and its use in outer space, only a portion of the components in the satellite can be repaired by very limited maintenance

TABLE 2.5

The Errors between the Simulation Method and the Approximation Method

Lifetime Distributions	Weibull	Uniform	Lognormal
Max error	2.2×10^{-3}	3.1×10^{-3}	1.19×10^{-3}
Mean error	6.31×10^{-4}	6.77×10^{-4}	4.70×10^{-4}
Calculation time of the approximation method (s)	2.21	2.31	1.568
Calculation time of the simulation method (s)	141.28	180.60	112.46

resources. To evaluate the reliability of this type of PMS, the use of a state-space model is necessary. In a traditional Markov chain, the sojourn time among states follows the exponential distribution [36]. But many real-world systems like the satellite consist of mechanical or electromechanical components whose lifetimes and repair times are very likely to follow non-exponential distributions such as the Weibull distribution. With the non-exponential lifetime distributions, the system cannot be modeled by the traditional Markov process. However, the SMP [23], belonging to the non-Markovain family, can deal with the non-exponential transition times. Therefore, the SMP, in conjunction with the modularization method and PMS-BDD models, is discussed in this chapter to evaluate the reliability of the complex PMS with partially repairable or non-repairable components.

In the man-made satellites, the altitude and orbit control system (AOCS) is a very critical subsystem. It is used to control and adjust the orbit and altitude for the whole lifetime. In this chapter, the use of AOCS is shown to illustrate the modeling process of PMS reliability. The AOCS consists of three subsystems—the control subsystem (Altitude and Orbit Control Computer), the sensor subsystem (including Sun Sensor, Earth Sensor, Star Track Sensor, and Gyro Assembly), and the actuator subsystem (Thrusters and Momentum Wheels). The AOCS can be regarded as a feedback system with three steps shown in Figure 2.9. In the first step, the sensor subsystem acquires and collects the position and altitude data. In the second step, the position and altitude data are transited to and analyzed by the control subsystem. In the third step, according to the analyzed result, the control subsystem sends orders to the actuator system to adjust the position and altitude. Next comes the measurement and adjustment procedure. The repetition of this procedure keeps the manned satellite in the right altitude and orbit for the whole lifetime.

As different tasks need to be completed in different phases, the whole lifetime of the AOCS can be divided into three phases—launching phase, orbit transfer phase, and orbital operation phase. In each phase, the system will execute different tasks in conjunction with other subsystems in the satellite.

The control subsystem consists of two components, computer A and cold standby computer B. Computer A can be repaired by backup components. The sensor subsystem consists of four components—sun sensor (C), earth

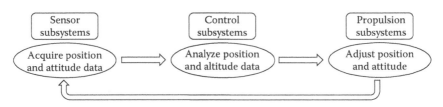

FIGURE 2.9
The working process of the satellite AOCS.

sensor (D), star track sensor (E), and gyro assembly (F). Only a portion of these sensors work in one specific phase. For example, the sun sensor and the earth sensor work in phase 1 and the earth sensor, the star track sensor, and the gyro assembly work in phase 2. The actuator subsystem consists of two parts—thrusters and momentum wheels which will work as actuators in different phases. There are two types of thrusters—two 15 N thrusters in cold standby (H and I, used for slightly adjustment during orbit transfer) and one 620 N thruster (G, the major power source for orbit transfer and adjustment). The momentum wheels (J, K, and L) are designed as a 2-out-of-3 system and they work in phase 3. The FT models for the three phases are shown in Figure 2.10. The parameters of all the components are listed in Table 2.6. The phase durations are $T_1 = 48$, $T_2 = 252$, and $T_3 = 50000$, respectively.

2.3.1.2 The Modular Approach

As can be seen in Figure 2.10, if the state-space model is directly applied, the FT models are too complicated to solve due to the state explosion problem. To address this problem, a modularization method defined in [37] and used in the PMS [13] is applied to deal with the problem. A phase module of a multi-phased system must meet two conditions: (1) a module is a set of basic events, which means a module must be a subset of all basic events; (2) for each phase, the basic events in the collection form an independent subtree in the FT. In other words, different subtrees in one phase should be independent of each other. The modularized fault tree (MFT) consists of independent subtrees

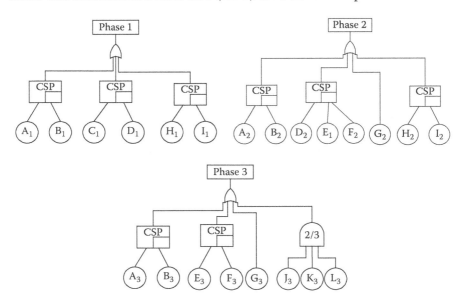

FIGURE 2.10
The FT model for three phases of the satellite AOCS.

TABLE 2.6
Parameters for the MSS in a Non-Markovian Environment

Components	Phase 1		Phase 2		Phase 3	
	α_1	β_1	α_2	β_2	α_3	β_3
A\B	2	500	2	500	1.5	8×10^4
AG	2	900	2	900	2	900
C\D\E\F	1.5	400	1.5	600	3	5×10^4
EG	1.8	600	1.8	600	1.8	600
G	1.5	1000	1.5	1500	2	5×10^4
H\I	2	600	1.8	300	1.5	1.5×10^4
J\K\L	1.5	1500	2.5	1500	2.5	1.1×10^4

(modules), and as a result, the complicated PMS can be assessed easily by the PMS-BDD method and module reliabilities.

The modular method can be applied in four steps for the reliability assessment of the PMS as follows:

- Step 1: Divide the FTs of the three phases into several independent subtrees by the modularization method. According to their own characteristics, the subtrees (modules) can be divided into static modules and dynamic modules. A module is a static module if it contains only static logic gates (OR, AND, or k-out-of-n). If there are dynamic logic gates, such as CSP, the module is a dynamic module [7].
- Step 2: After modularization, the modules can be treated as the bottom events of the MFTs. In the MFT, the modules are independent of each other.
- Step 3: According to the characteristics of the modules, the reliability indices of dynamic and static modules can be evaluated by SMP as well as the approximation method and mini-component method, respectively.
- Step 4: Integrating the results of step 2 and step 3, the system reliability can be assessed by using the PMS-BDD method.

There are 12 basic events $\{A, B, C, D, E, F, G, H, I, J, K, L\}$ in this PMS. All the components can be divided into several modules in the three phases. The relationship between independent modules and basic events in three phases is shown as

$$\pi_1 = \{M_{1,1} = (A, B), M_{2,1} = (C, D), M_{4,1} = (H, I)\}$$

$$\pi_2 = \{M_{1,2} = (A, B), M_{2,2} = (D, E, F), M_{3,2} = G, M_{4,2} = (H, I)\}$$

$$\pi_3 = \{M_{1,3} = (A, B), M_{2,3} = (E, F), M_{3,3} = G, M_{5,3} = (J, K, L)\}$$

where π_i, $i = 1, 2, 3$ is the set of components working in phase i and $M_{i,j}$ is the ith module in phase j. Because some modules (e.g., M_2) are not consistent across phases, the modules across phases need to be formed. With the set theory, the modules across phases can be obtained and shown as [3]

$$\{M_1 = (A, B), M_2 = (C, D, E, F), M_3 = G, M_4 = (H, I), M5 = \{J, K, L\}\}. \quad (2.17)$$

After the system modularization, all the basic events of the original FT model are divided into five independent modules. All the modules can be treated as basic events in the MFT model, as shown in Figure 2.11.

2.3.1.3 Module and System Reliability

2.3.1.3.1 The Static Module

In this section, the mini-component method is used to evaluate the reliability of the static module. In the MFT, module 3 and module 5 are static modules. Here, we use module 5 as an example. Module 5 consists of three components, and system operation requires that at least two components be operational. The failure probability of a k-out-of-n system such as module 5 (2-out-of-3 system) in a single phase can be expressed as [38]

$$p_{M_5}(t) = \sum_{i=k}^{n} C_n^i \left(1 - F_J(t)\right)^i \left(F_J(t)\right)^{n-i}, \quad F_J(t) = 1 - e^{-(t/\beta_J)^{\alpha_J}} \quad (2.18)$$

where $F_J(t)$ is the CDF of the failure probability in module 5.

To deal with the dependence among static modules, a set of mini-components [2] is used to replace the unit in one specific phase. With the

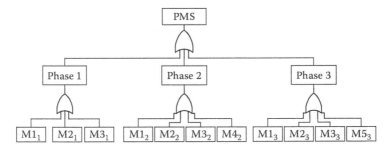

FIGURE 2.11
The modularized FT (MFT) model of the satellite AOCS. (a) Transition diagram of module 2 in phase 1. (b) Transition diagram of module 2 in phase 2. (c) Transition diagram of module 2 in phase 3.

Reliability Modeling of PMS

mini-components method, the CDF of module 5 in phase j—$F_{M_2,j}(t)$ can be expressed as

$$F_{M_2,j}(t) = \left[1 - \prod_{i=1}^{j-1}\left(1 - p_{M_2,i}(T_i)\right)\right] + \left[\prod_{i=1}^{j-1}\left(1 - p_{M_2,i}(T_i)\right)\right] \cdot p_{M_2,i}(t) \quad (2.19)$$

where T_i is the time duration of phase i and $p_{M_2,j}(t)$ is the failure probability of unit M_2 at time t. Time t is measured from phase j. The first term of Eq. 2.14 is the probability that the system fails in the first $j-1$ phases and the second term is the probability that the system fails at time t in phase j.

With the parameters shown in Table 2.6 and Eqs. 2.18 and 2.19, the reliability of module 5 at the end of each phase, $R_{M_5,j}(t)$, can be evaluated, and the results are shown in Table 2.7. Similarly, the reliability of module 3 can also be evaluated, and the results are also shown in Table 2.7.

2.3.1.3.2 The Dynamic Module without Structure Variation

From Figure 2.10, we can see that modules M_1 and M_4 are dynamic modules without structure variation. In other words, the system structures of these two modules do not change in the whole lifetime. The module state probabilities at different times in one phase can be evaluated by the approximation method proposed in Section 2.2.1. To deal with the dependence across phases, the module state of phase i is set to be equal to the state at the end of last phase. By this method, the state probabilities of module 1 and module 4 are evaluated, and are shown in Tables 2.8 and 2.9, respectively.

TABLE 2.7

Parameters for the MSS in a Non-Markovian Environment

	Phase 1	Phase 2	Phase 3
$R_{M_5,j}$	N/A	N/A	0.9537
$R_{M_3,j}$	N/A	0.9335	0.9098

TABLE 2.8

State Probabilities of Module 1

	S1	S2	S3	S4
T_1	0.9999	1.7669×10^{-4}	1.2071×10^{-10}	1.132×10^{-8}
T_2	0.9961	0.0039	1.4015×10^{-9}	3.2951×10^{-6}
T_3	0.9795	0.0203	1.459×10^{-7}	1.4861×10^{-4}

TABLE 2.9

State Probabilities of Module 4

	S1	S2	S3
T_1	0.9999	6.1359×10^{-5}	6.2723×10^{-10}
T_2	0.9900	0.0099	1.7252×10^{-4}

2.3.1.3.3 The Dynamic Module with Structure Variation

From the FT model of module 2, we can see that the components working in different phases are different. Therefore, a specially designed method is needed to deal with module 2. It is important to note that although the system structures are different in different phases, the components' states are the same between the end of one phase and the beginning of the next phase. Based on this fact, the module reliability can be assessed by three steps:

- Step 1: Construct the state transition diagrams for each phase according to their dynamic behaviors. In this section, module 2 components are working in three phases, and the state transition diagram for each phase are shown in Figure 2.12.

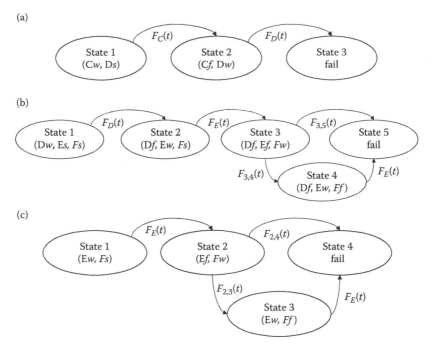

FIGURE 2.12
The state transition diagram for module 2 in three phases.

Reliability Modeling of PMS

- Step 2: Construct the relationship between the states of every two adjacent phases according to the system structure [5]. The relationship is shown in Figure 2.13. From Figure 2.13, we can see that state 1 and state 2 of phase 1 are mapped into state 1 of phase 2. The reason is that component D does not fail in either state in phase 1. Module 2 will stay in state 1 at the beginning of phase 2 if component D does not fail in phase 1.
- Step 3: Evaluate the module reliability by the approximation method from one phase to the next according to the relationship between every two adjacent phases and the approximation method. The module state probabilities at the end of each phase are shown in Table 2.10.

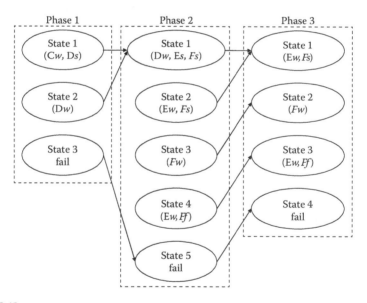

FIGURE 2.13
The states relationship between two adjacent phases of module 2.

TABLE 2.10
State Probabilities of Module 2

	S1	S2	S3	S4	S5
T_1	0.9987	0.0013	4.4752×10^{-6}	N/A	N/A
T_2	0.9889	0.0110	3.6160×10^{-5}	7.7974×10^{-9}	1.0517×10^{-12}
T_3	0.8617	0.1372	1.8283×10^{-4}	1.5244×10^{-4}	N/A

2.3.1.3.4 System Reliability

In this section, it will be shown that with the independent basic events, the system reliability can be assessed by the efficient 5-step PMS-BDD method proposed by Zang and Trivedi [2]. The PMS-BDD method can combine the BDD models of different phases by phase algebra shown in Table 2.11 to obtain the system BDD to evaluate the system reliability. If the variables linked by edges directly belong to different variables, the evaluation method will be the same as the traditional BDD method. But if they belong to the same components in different phases, the phase algebra can be used to decrease amount of the final BDD model.

On the other hand, the size of a BDD heavily depends on the order of variables. There exist two phase-dependent operation (PDO) ways—forward PDO and backward PDO. According to [2], the BDD generated by backward PDO is much smaller than that generated by the forward PDO so that computation in the last step is much easier. Applying the backward PDO in the AOCS and taking an order of $M_{1,3} < M_{1,2} < M_{1,1} < M_{2,3} < M_{2,2} < M_{2,2} < M_{3,3} < M_{3,2} < M_{3,1} < M_{4,2} < M_{5,3}$, the system BDD model can be transferred from the system MFT model, as shown in Figure 2.14.

The system reliability R_{sys} is the probability of the SDP from the root to the vertex "0" through the system BDD figure. From Figure 2.14, we can get the disjoint path: $M_{1,3}M_{2,3}M_{3,3}M_{4,2}M_{5,3}$. According to the SDP, the system reliability can be assessed by

TABLE 2.11

Rules of Phase Algebra ($i < j$)

$M_i \cdot M_j \to M_j$	$\overline{M_i} + \overline{M_j} \to \overline{M_j}$
$\overline{M_i} \cdot \overline{M_j} \to \overline{M_i}$	$M_i + M_j \to M_i$
$\overline{M_i} \cdot M_j \to 0$	$\overline{M_i} + M_j \to 1$

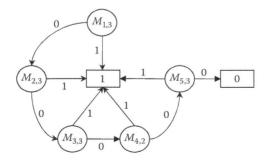

FIGURE 2.14
The modularized BDD model for the satellite AOCS.

$$R_{sys} = P\left(\overline{M_{1,3}}\,\overline{M_{2,3}}\,\overline{M_{3,3}}\,\overline{M_{4,2}}\,\overline{M_{5,3}}\right)$$

$$= P\left(\overline{M_{1,3}}\right) P\left(\overline{M_{2,3}}\right) P\left(\overline{M_{3,3}}\right) P\left(\overline{M_{4,2}}\right) P\left(\overline{M_{5,3}}\right)$$

$$= P\left(1 - P\left(S5_{M1}(T_3)\right)\right) \cdot P\left(1 - P\left(S6_{M2}(T_3)\right)\right) \cdot R_{M3}(T_w)$$

$$\times P\left(1 - P\left(S3_{M4}(T_3)\right)\right) \cdot R_{M5}(T_w) \tag{2.20}$$

where Si_{Mj} represents the state of module j and T_w represents the whole lifetime, $T_w = T_1 + T_2 + T_3$. With Eq. 2.20, the system reliability of the PMS can be computed as 0.8804394.

2.3.2 PMS Analysis Considering Random Shocks Effect

PMSs are widely used, especially in the aerospace industry. In most of their lifetime, the aerospace equipment stays in the outer space. There are many kinds of cosmic rays in outer space, such as the Galactic Cosmic Rays, which randomly hit these systems and cause significant impact on the electronics inside or outside the equipment. For example, the ionizing nature of GCR particles can pose significant threats to the electronics located onboard, such as the microprocessors to which they may cause memory bit flips and latch-ups. This kind of phenomenon is generally called the single event effect (SSE) [39] and occurs randomly, that is, as a random shock. If these random shocks are not considered, the reliability of the PMS will be overestimated.

Random shocks have been examined with different approaches in reliability modeling. Esary [40] studied extreme shocks in components' reliability assessment. Lin and Zio [16] studied the components' reliability considering both degradation processes and random shocks. At the system level, Wang and Pham [41] investigated systems subject to degradation and random shocks, in which the random shocks can lead the system to fail immediately. Rafiee [42] studied cumulative random shocks, which increase the components' failure rates. Berker [43] used a semi-Markov model to describe a system subject to random shocks. Recently, Ruiz-Castro [44] studied a system subject to external shocks leading to extreme failures and cumulative damage.

In this part, the MRGP is used to describe the hybrid components' lifetime distributions and the dynamic behaviors in the PMSs. The reliability of PMSs subject to random shocks is, then, evaluated by the MC simulation method.

2.3.2.1 MRGP and Multistate Random Shocks Model

Though the SMP can deal with the non-exponential distributions, it is not available in some situations, for example, a cold standby system with two working components (H, I) and one switch component (S) whose state transition graph is shown in Figure 2.15. Since the distribution $F_H(t)$ from state S1 to state S2 is different from the distribution $F_H(t)$ from state S3 to state S5 as component H is not regenerated in state S3, the system cannot be modeled by an SMP. The use of MRGP to deal with this problem is shown in this section.

2.3.2.1.1 Basic Conceptions of the MRGP

A stochastic process is an MRGP $\{Z_t, t \geq 0\}$ if it exhibits an embedded MRS $\{X, S\}$ with the additional property that all conditional finite distributions $\{Z_{t+S_n}, t \geq 0\}$ given $\{Z_u; 0 \leq u \leq S_n, X_n = i, i \in \Omega\}$ are the same as those of $\{Z_t, t \geq 0\}$ given $X_0 = i$ [17],

$$\Pr\{Z_{t+S_n} = j \mid 0 \leq u \leq S_n, X_n = i\} = \Pr\{Z_t = j \mid X_0 = i\}. \quad (2.21)$$

To evaluate the state probabilities of the MRGP, the same variables used in the SMP model in the previous section such as $\boldsymbol{\theta}(t) = V_{i,j}(t)$ representing the conditional transition probabilities and $\boldsymbol{Q}(t) = Q_{i,j}(t)$ representing the one-step transition probabilities are used in the MRGP model. For the MRGP, the Markov renewal equation is different from the SMP, which is shown as follows:

$$\theta_{i,j}(t) = E_{i,j}(t) + \sum_{k=1}^{K} \int_0^t q_{i,k}(\tau) \theta_{k,j}(t-\tau) d\tau \quad (2.22)$$

$$E_{i,j}(t) = \Pr\{Z_t = j, S_1 > t \mid X_0 = i\} \quad (2.23)$$

where $E_{i,j}(t)$ is the probability that the system state changes from state i to state j before the first regeneration.

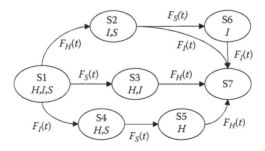

FIGURE 2.15
The state transition diagram for the cold standby system example.

2.3.2.1.2 The Multistate System Model Considering Random Shocks

As described in the previous section, the randomly occurring cosmic rays affect the electronics as random shocks. To integrate the random shocks in the PMS reliability model, some basic assumptions are made:

- The arrival of the random shocks follow a homogeneous Poisson process [16], with a constant arrival rate u_n in phase n (shown as Figure 2.16); the arrival rate may change from phase to phase, due to the change of environment in the different phases.
- The random shocks are s-independent of the components' failure process.
- The damage brought by the random shocks is cumulative; specifically, the random shocks increase the failure rate of a constant amount ε each time the shocks occur and do not lead the components to failure directly.

We assume that M indicates the system state and N indicates the number of random shocks that have occurred in this section. To integrate the random shocks in the PMS reliability model, the system state indicator is extended from M into (M, N). Using the cold standby system shown in Figure 2.15 as an example, the random shocks are integrated into the cold standby system. The state transition diagram for the system considering random shocks for M_3 is shown in Figure 2.17. Furthermore, the failure rates after n random shocks $\lambda_{i,j}^n$ in Figure 2.17 are set to $\lambda_{i,j}^n = \lambda_{i,j}(1+\varepsilon)^n$, where $\lambda_{i,j}$ represents the transition rate of the system from state i to state j without random shocks and $(1+\varepsilon)^n$ characterizes the cumulative effect of the random shocks. In the case study of this section, the failure rate increment ε, due to a shock, is set to 0.3 [45]. The MC simulation method is shown to be used to assess the PMS reliability in this section. A simulation procedure is illustrated in the next section.

2.3.2.2 Simulation Procedure

During each simulation of the system phase by phase, three quantities are recorded—module state M, number of random shocks N, system failure time T_F, the lifetimes of the non-exponential components T_{NE}, the time of the system has been through before next regeneration point T_{epoch}. Using the cold standby system in Figure 2.17 as an example, the initial system state vector

FIGURE 2.16
The random shocks process in phase n.

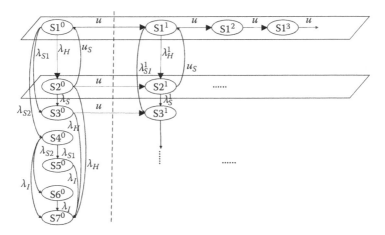

FIGURE 2.17
The state transition diagram with random shocks for the example cold standby system.

is set to be $(M_0, N_0, T_{F,0}, \boldsymbol{T}_{NE}, T_{epoch}) = (7, 0, 0, (0,0), 0)$. In each repeated simulation history, the failure time of the system T_F is sampled from the beginning of the first phase to the end of the last phase and it is recorded as the outcome at the end of the last phase. The simulation of the module in each phase begins by reading the system vector from the initial system state vector or the system state vector from the previous phase. The system simulation procedure is shown in Figure 2.18.

The *phase simulation procedure* in Figure 2.18 for phase *n* is as follows:

Set $S = (M_n, N_n, T_{F,n}, \boldsymbol{T}_{NE}, T_{epoch})$
Set phase working time $T = 0$
While $T \leq T_n$ (T_n is the working time of phase N), do the following.

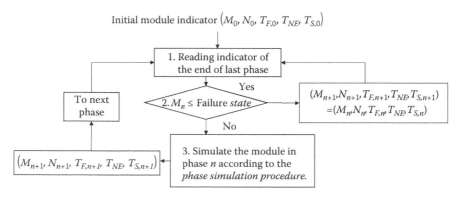

FIGURE 2.18
The system simulation procedure for the PMS under random shocks.

1. If $n = 1$, sample the transition time vector of the non-exponential components T_{NE}.
2. Calculate the system state transition parameter based on N_n.
3. Sample the transition time vector $\mathbf{X}$.
    1. **If** x_j follows the exponential distribution, simulate x_j and calculate $x_j = x_j + T_{epoch}$.
    2. **Else if** x_j follows the non-exponential distribution, and set x_j equal to the corresponding element in T_{NE} according to their own behavior.

    End if
    3. Simulate the random shock occurrence time y.
4. Compare x_j and y.
    If x_j is the smallest, the system moves to $S = (j, N_n, T_{F,n-1})$, $t' = x_j$.
    If x_j is exponential, set $T_{epoch} = T_{epoch} + x_j$ (the system state changes before the regeneration epoch).
    Else if x_j is non-exponential, set $T_{epoch} = 0$ (the system state changes at the regeneration epoch).

    End if
    Else if y is the smallest, the system state moves to $S = (M_n, N_n + 1, T_{F,n-1})$, $t' = y$,
        Set $T_{epoch} = T_{epoch} + y$.
    End if
5. Set $T = T + t'$.
6. **If** $S(1) \leq$ *Failure State*
    Then break.
    End if

End While

2.3.2.3 Case Study

2.3.2.3.1 System Structure

In this section, an AOCS of the manned spacecraft is applied and it is slightly different from the AOCS in the previous section. According to the different missions that need to be accomplished, the whole lifetime of the AOCS can be divided into four phases: launching phase, orbit-transfer phase, on-orbit phase, and back to earth phase. In our work, we assume that the phase durations are $T_1 = 36$ h, $T_2 = 240$ h, $T_3 = 960$ h, and $T_4 = 36$ h, respectively. The FT models for the four phases are shown in Figure 2.19.

The spacecraft AOCS is composed of three functional parts; (1) microcomputers (processors): Computers (A and B) and functional part (C). (2) Sensors: sun sensor (D), Earth sensor (E), star track sensor (F), and gyro

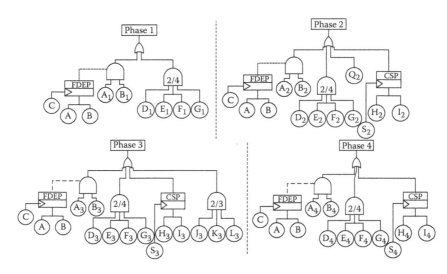

FIGURE 2.19
The FT models for each phase of the spacecraft AOCS.

assembly (G). (3) Actuators: (1) thruster 1 (20 N, cold standby subsystem thrusters H, I, and switch component S); (2) thruster 2 (620 N, Q); (3) three momentum wheels (2 out of 3 subsystems, J, K, L).

All the components can be divided into two categories: mechanical components and electrical components. The commonly seen failure modes of the mechanical components are fracture, fatigue, or corrosion [46,47], which cannot be maintained without backup standby. For the electronics, except for the structure failures (that also cannot be maintained without backup standby), they are subject to functional failures (like the computer crash or power loss), which can be self-repaired by system reboot [48]. The lifetime of the mechanical components is usually described by a Weibull distribution, whereas the lifetime of the electronics is described by the exponential distribution. All the components' parameters are shown in Table 2.12.

TABLE 2.12

The Parameters for the AOCS in the Spacecraft

	A	B	G	H	I	Q		
α	2×10^4	3×10^4	3×10^4	3×10^4	2×10^4	2×10^4		
β	2	1.5	1.5	2	1.5	3.5		
	C	D	E	F	J	K	L	S1	S2
λ	1/20000	1/30000	1/30000	1/30000	1.5×10^4	1.5×10^4	1.5×10^4	1/20000	1/30000
u	1/20000	1/20000	1/20000	1/20000	1/20000	1/20000	1/20000	1/20000	

In this example, the system FT model is also complicated to solve, so the modular method is applied. The system can be divided into five independent modules: $M1 = (A,B,C)$, $M2 = (D,E,F,G)$, $M3 = (H,I,S)$, $M4 = J$, $M5 = (J,K,L)$. With these independent modules, the system FT models can be simplified and shown in Figure 2.20.

According to the above analysis, the reliability of the PMS considering random shocks can be assessed by the following steps: (1) system modularization to simplify the system FT models and to obtain modules independent of each other in the MFT; (2) integration of the random shocks into the PMS and evaluation of the reliability indices of all the modules by SMP modeling and MC simulation; (3) combination of the MFT and assessment of the reliabilities of the independent modules, to assess the reliability of the PMS.

2.3.2.3.2 Results and Analysis

According to the system FT model, the MC simulation procedure, and system reliability analysis method shown in Section 2.3.1.3, the system reliability of the PMS subject to random shocks can be evaluated. The system reliability of the phased AOCS considering random shocks is shown as the dotted line in Figure 2.21. Moreover, the reliability of the same system without the random shocks is shown as the solid line in Figure 2.21. The reliabilities of the AOCS at the end of each phase are also shown in Table 2.13, as well as the relative difference between the reliabilities with and without random shocks and the average number of shocks occurred in each phase.

As expected, the system reliability is lower when considering random shocks, especially in phase 3 and phase 4, when the system travels a long time in the outer space and after. If the random shocks are not considered in the modeling, the system reliability may be overestimated.

With respect to the random shocks modeling, we have analyzed the sensitivity of the system reliability estimates for two parameters, the random

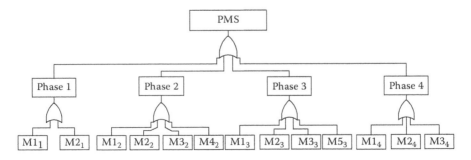

FIGURE 2.20
The MFT of the spacecraft AOCS.

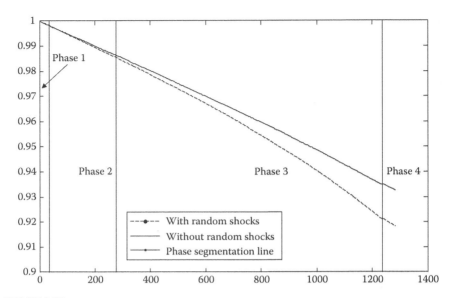

FIGURE 2.21
The reliability of the spacecraft AOCS with and without random shocks.

TABLE 2.13

The Results for the Spacecraft AOCS with and without Random Shocks

	Phase 1	Phase 2	Phase 3	Phase 4
$R_{sys}^{noshock}(t)$	0.9982	0.9862	0.9340	0.9317
$R_{sys}^{shocks}(t)$	0.9981	0.9853	0.9214	0.9184
Relative difference	1.0018e−04	9.1259e−04	0.0135	0.0143
Average number of shocks	0.0355	0.2422	0.9599	0.0484

shocks occurrence rate $u = [1/500, 1/800]\,h^{-1}$ and the relative increment in the transition rates $\varepsilon = [0.2, 0.5]$. The estimated system reliabilities for different combinations of the two parameters are shown in Figure 2.22.

From Figure 2.22 we can see that with the increase of the relative increment ε or the random shocks occurrence rate u, the system reliability decreases as expected. The higher ε leads to larger components' failure rates, and larger occurrence rate u values result in more random shocks over the whole lifetime, which decreases the system reliability. In Table 2.14, the system reliability with parameters $\varepsilon = 0.2$ and $u = 1/800\,h^{-1}$ is set to the standard and other elements are differences between the standard and the results by different parameters combination. From Table 2.14, we can see that when the same percentage of variation is applied to the two parameters, ε is more influential than u on the system reliability.

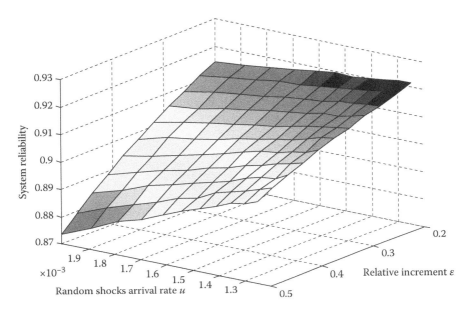

FIGURE 2.22
The reliability of the spacecraft AOCS with and without random shocks for different combinations of u and ε.

2.4 Conclusion

This chapter presents two approaches of the evaluation methods to the non-exponential dynamic system and their use in the reliability analysis of the PMS. Traditionally, the commonly seen approaches of the PMS analysis methods are the combinatorial methods or the state-space model. The combinatorial methods are computationally efficient but not available in the dynamic system. The state-space-based methods can deal with various dynamic systems but suffer from the state explosion problem. The modular method is much better and combines the advantages of both methods.

TABLE 2.14
The Errors of the Results for the AOCS Sensitivity Analysis

System Reliability	0.2	0.26	0.32	0.38	0.44	0.5
0.00200	0%	0.12%	0.25%	0.33%	0.55%	0.76%
0.00179	0.47%	0.55%	0.73%	0.86%	1.11%	1.47%
0.00161	0.87%	1.0%	1.26%	1.49%	1.77%	2.21%
0.00147	1.27%	1.54%	1.77%	2.14%	2.58%	3.05%
0.00135	1.77%	2.07%	2.49%	2.85%	3.30%	4.03%
0.00125	2.35%	2.61%	3.08%	3.61%	4.22%	5.22%

On the other hand, the non-exponential distribution is more practical to describe the lifetime of the components, but the traditional Markov model is not capable of doing so. Therefore, the use of the SMP or the MRGP that belong to the Markov renewal theory is shown in this chapter to deal with this problem. In this chapter, the reliability of the PMS with partially repairable components is evaluated in detail. Furthermore, the PMSs subject to random shocks is also discussed.

References

1. Xing, L. Reliability evaluation of phased-mission systems with imperfect fault coverage and common-cause failures. *IEEE Transactions on Reliability*, 2007, 56(1): 58–68.
2. Zang, X., Sun, N., Trivedi, K.S. A BDD-based algorithm for reliability analysis of phased-mission systems. *IEEE Transactions on Reliability*, 1999, 48(1): 50–60.
3. Xing, L., Dugan, J.B. Analysis of generalized phased-mission system reliability, performance, and sensitivity. *IEEE Transactions on Reliability*, 2002, 51(2): 199–211.
4. Levitin, G., Xing, L., Amari, S.V. Recursive algorithm for reliability evaluation of non-repairable phased mission systems with binary elements. *IEEE Transactions on Reliability*, 2012, 61(2): 533–542.
5. Wang, C., Xing, L., Peng, R. et al. Competing failure analysis in phased-mission systems with multiple functional dependence groups. *Reliability Engineering & System Safety*, 2017, 164: 24–33.
6. Xing, L., Amari, S.V. Chapter 23: Reliability of phased-mission systems. In *Handbook of Performability Engineering*, K.B. Misra, Ed. Berlin: Springer, 2008: 349–368.
7. Li, Y.F., Mi, J., Liu, Y. et al. Dynamic fault tree analysis based on continuous-time Bayesian networks under fuzzy numbers. Proceedings of the Institution of Mechanical Engineers, Part O. *Journal of Risk and Reliability*, 2015, 229(6): 530–541.
8. Xing, L., Levitin, G. BDD-based reliability evaluation of phased-mission systems with internal/external common-cause failures. *Reliability Engineering & System Safety*, 2013, 112: 145–153.
9. Tang, Z., Dugan, J.B. BDD-based reliability analysis of phased-mission systems with multimode failures. *IEEE Transactions on Reliability*, 2006, 55(2): 350–360.
10. Mo, Y., Xing, L., Dugan, J.B. MDD-based method for efficient analysis on phased-mission systems with multimode failures. *IEEE Transactions on Systems, Man, and Cybernetics: Systems*, 2014, 44(6): 757–769.
11. Alam, M., Al-Saggaf, U.M. Quantitative reliability evaluation of repairable phased-mission systems using Markov approach. *IEEE Transactions on Reliability*, 1986, 35(5): 498–503.
12. Mura, I., Bondavalli, A. Markov regenerative stochastic Petri nets to model and evaluate phased mission systems dependability. *IEEE Transactions on Computers*, 2001, 50(12): 1337–1351.

13. Ou, Y., Dugan, J.B. Modular solution of dynamic multi-phase systems. *IEEE Transactions on Reliability*, 2004, 53(4): 499–508.
14. Meshkat, L., Xing, L., Donohue, S.K. et al. An overview of the phase-modular fault tree approach to phased mission system analysis. Space Mission Challenges for Information Technology, http://hdl.handle.net/2014/, 2003.
15. Liu, Y., Chen, C.J. Dynamic reliability assessment for nonrepairable multistate systems by aggregating multilevel imperfect inspection data. *IEEE Transactions on Reliability*, 2017, 66(2): 281–297.
16. Lin, Y.H., Li, Y.F., Zio, E. Integrating random shocks into multi-state physics models of degradation processes for component reliability assessment. *IEEE Transactions on Reliability*, 2015, 64(1): 154–166.
17. Bai, D.S., Chung, S.W., Chun, Y.R. Optimal design of partially accelerated life tests for the lognormal distribution under type I censoring. *Reliability Engineering & System Safety*, 1993, 40(1): 85–92.
18. Telek, M., Pfening, A. Performance analysis of Markov regenerative reward models. *Performance Evaluation*, 1996, 27: 1–18.
19. Limnios, N., Oprisan, G. *Semi-Markov Processes and Reliability*. New York: Springer Science & Business Media, 2012.
20. Janssen, J., Manca, R. *Semi-Markov Risk Models for Finance, Insurance and Reliability*. New York: Springer Science & Business Media, 2007.
21. Yin, L., Fricks, R.M., Trivedi, K.S. Application of semi-Markov process and CTMC to evaluation of UPS system availability. In *Proceedings of the Annual Reliability and Maintainability Symposium, 2002*. IEEE, Seattle, 2002: 584–591.
22. Pievatolo, A., Tironi, E., Valade, I. Semi-Markov processes for power system reliability assessment with application to uninterruptible power supply. *IEEE Transactions on Power Systems*, 2004, 19(3): 1326–1333.
23. Distefano, S., Trivedi, K.S. Non-Markovian state-space models in dependability evaluation. *Quality and Reliability Engineering International*, 2013, 29(2): 225–239.
24. Lisnianski, A., Levitin, G. *Multi-State System Reliability Assessment, Optimization and Applications*. Singapore: World Scientific Publishing, 2003.
25. Lisnianski, A., Frenkel, I., Ding, Y. *Multi-State System Reliability Analysis and Optimization for Engineers and Industrial Managers*. London: Springer, 2010.
26. Wu, X., Hillston, J. Mission reliability of semi-Markov systems under generalized operational time requirements. *Reliability Engineering & System Safety*, 2015, 140: 122–129.
27. Csenki, A. An integral equation approach to the interval reliability of systems modelled by finite semi-Markov processes. *Reliability Engineering & System Safety*, 1995, 47(1): 37–45.
28. Boehme, T.K., Preuss, W., Van der Wall, V. On a simple numerical method for computing Stieltjes integrals in reliability theory. *Probability in the Engineering and Informational Sciences*, 1991, 5(1): 113–128.
29. Zio, E., Pedroni, N. Reliability estimation by advanced Monte Carlo simulation. In Javier Faulin, Angel A. Juan, Sebastián Salvador Martorell Alsina, Jose Emmanuel Ramirez-Marquez (Eds.), In *Simulation Methods for Reliability and Availability of Complex Systems*. London: Springer, 2010: 3–39.
30. Zio, E., Librizzi, M. Direct Monte Carlo simulation for the reliability assessment of a space propulsion system phased mission (PSAM-0067). Proceedings of the *Eighth International Conference on Probabilistic Safety Assessment & Management (PSAM)*. New York, USA: ASME Press, 2006.

31. Gillespie, D.T. Monte Carlo simulation of random walks with residence time dependent transition probability rates. *Journal of Computational Physics*, 1978, 28(3): 395–407.
32. Castet, J.F., Saleh, J.H. Satellite and satellite subsystems reliability: Statistical data analysis and modeling. *Reliability Engineering & System Safety*, 2009, 94(11): 1718–1728.
33. Castet, J.F, Saleh, J.H. Satellite reliability: Statistical data analysis and modeling. *Journal of Spacecraft and Rockets*, 2009, 46(5): 1065.
34. Saleh, J.H., Castet, J.F. *Spacecraft Reliability and Multi-state Failures: A Statistical Approach*. Hoboken, New Jersey: John Wiley & Sons, 2011.
35. Peng, W., Li, Y.F., Yang, Y.J. et al. Leveraging degradation testing and condition monitoring for field reliability analysis with time-varying operating missions. *IEEE Transactions on Reliability*, 2015, 64(4): 1367–1382.
36. Coit, D.W. Cold-standby redundancy optimization for nonrepairable systems. *IIE Transactions on Reliability*, 2001, 33(6): 471–478.
37. Levitin, G., Xing, L., Dai, Y. Minimum mission cost cold-standby sequencing in non-repairable multi-phase systems. *IEEE Transactions on Reliability*, 2014, 63(1): 251–258.
38. Dutuit, Y., Rauzy, A. A linear-time algorithm to find modules of fault trees. *IEEE Transactions on Reliability*, 1996, 45(3): 422–425.
39. Huang, J., Zuo, M.J. Multi-state k-out-of-n system model and its applications. *Proceedings of the Annual Reliability and Maintainability Symposium*, 2010: 264–268.
40. Golge, S., O'Neill, P.M., Slaba, T.C. NASA Galactic Cosmic Radiation Environment Model: Badhwar-O'Neill, 34th *International Cosmic Ray Conference, 30 Jul.-6 Aug. 2015, The Hague, Nethelands*, 2015.
41. Esary, J.D., Marshall, A.W. Shock models and wear processes. *The Annals of Probability*, 1973: 627–649.
42. Li, W., Pham, H. Reliability modeling of multi-state degraded systems with multi-competing failures and random shocks. *IEEE Transactions on Reliability*, 2005, 54(2): 297–303.
43. Rafiee, K., Feng, Q., Coit, D.W. Reliability modeling for dependent competing failure processes with changing degradation rate. *IIE Transactions*, 2014, 46(5): 483–496.
44. Becker, G., Camarinopoulos, L., Kabranis, D. Dynamic reliability under random shocks. *Reliability Engineering & System Safety*, 2002, 77(3): 239–251.
45. Ruiz-Castro, J.E. Markov counting and reward processes for analysing the performance of a complex system subject to random inspections. *Reliability Engineering & System Safety*, 2016, 145: 155–168.
46. Fan, J., Ghurye, S.G., Levine, R.A. Multicomponent lifetime distributions in the presence of ageing. *Journal of Applied Probability*, 2000, 37(2): 521–533.
47. Collins, J.A. *Failure of Materials in Mechanical Design: Analysis, Prediction, Prevention*. Hoboken, New Jersey: John Wiley & Sons, 1993.
48. Vigander, S. Evolutionary Fault Repair of Electronics in Space Applications, Dissertation. Norwegian University Science and Technology, Trondheim, Norway, Norway, Norwegian University Sci. Tech, February 28, 2001.

3

Optimal Periodic Software Rejuvenation Policies in Discrete Time—Survey and Applications

Tadashi Dohi, Junjun Zheng, and Hiroyuki Okamura
Hiroshima University

CONTENTS

- 3.1 Introduction 82
- 3.2 Model Description 84
  - 3.2.1 Notation 84
  - 3.2.2 Model Description 85
- 3.3 Expected Cost Analysis 87
  - 3.3.1 Formulation 87
  - 3.3.2 Statistical Inference 89
- 3.4 Availability Analysis 92
  - 3.4.1 Formulation 92
  - 3.4.2 Statistical Inference 93
- 3.5 Cost-Effectiveness Analysis 94
  - 3.5.1 Formulation 94
  - 3.5.2 Statistical Inference 95
- 3.6 Expected Total Discounted Cost Analysis 95
  - 3.6.1 Formulation 95
  - 3.6.2 Statistical Inference 98
- 3.7 Numerical Examples 99
  - 3.7.1 Case of Expected Cost 100
  - 3.7.2 Case of System Availability 101
  - 3.7.3 Case of Cost-Effectiveness 102
  - 3.7.4 Case of Expected Total Discounted Cost 103
- 3.8 Conclusions 104
- References 105

3.1 Introduction

Present-day applications in computer systems impose stringent requirements in terms of software dependability, because system failure, caused by software failure in almost all cases, may lead to a huge economic loss or risk to human life. A guaranteed fulfillment of these requirements is very difficult, especially in applications with nontrivial complexity. In recent years, considerable attention has been paid to continuously running software systems whose performance characteristics are smoothly degrading in time. When a software application executes continuously for a long period of time, some of the faults cause software to age due to the error conditions that accrue with time and/or load. This phenomenon is called *software aging* and can be observed in many original software systems [1–6]. One common experience suggests that most software failures are transient in nature [7]. Since transient failures disappear if the operation is retried later in slightly different context, it is difficult to characterize their root origin. Therefore, the residual software faults are obvious in the operational phase. Grottke and Trivedi [8] classify several software bugs and point out that the resource exhaustion in computer systems causes the software aging. A complementary approach to handle transient software failures is called *software rejuvenation* [9] which can be regarded as a preventive and proactive solution that is particularly useful for counteracting the phenomenon of software aging. It involves stopping the running software occasionally, cleaning its internal state, and restarting it. Cleaning the internal state of software may involve garbage collection, flushing operating system kernel tables, reinitializing internal data structures, etc. An extreme, but well-known example of rejuvenation is a hardware reboot. In this way, software rejuvenation is becoming much popular as one of the light weighted software fault tolerant techniques.

Huang et al. [9] propose a continuous time Markov chain (CTMC) with four states, that is, initial robust (clean), failure probable, rejuvenation, and failure states. They evaluate both the unavailability and the operating cost in the steady state under a random software rejuvenation schedule. Danjou et al. [10] and Dohi et al. [11–15] extend the result of Huang et al. [9] and propose different software rejuvenation models in continuous time, based on a semi-Markov process. Furthermore, Garg et al. [16], Suzuki et al. [17], and Dohi et al. [18] introduce the *periodic* rejuvenation and develop a Markov regenerative process (MRGP) model to trigger software rejuvenation on computer clock. In the above works, the authors propose nonparametric estimation algorithms with the empirical distribution to estimate the optimal software rejuvenation schedule statistically from the complete sample of failure time data. If a sufficient number of samples of failure time data can be obtained, then the estimates of the optimal software rejuvenation schedules based on Dohi et al.'s algorithm [11–14] asymptotically converge to the real optimal solutions. Zhao et al. [19] apply an accelerated life testing technique by injecting memory leaks in their experiments and examine the above nonparametric estimation methods in importance sampling simulation. Rinsaka and Dohi [20] propose another nonparametric estimation algorithm based on the kernel density estimation (see, e.g., Duin [21], Parzen [22], Silverman [23]) to improve the estimation accuracy of the optimal software rejuvenation schedule with a small sample data. Rinsaka and Dohi [24–26] use a nonparametric predictive inference (NPI) approach provided by Coolen and Yan [27] and Coolen-Schrijner and Coolen [28]. Though the NPI-based approach is categorized as a prediction-based software rejuvenation, the results obtained by Rinsaka and Dohi [24–26] are quite different from those of Vaidyanathan and Trivedi [29] for another prediction-based scheme with linear regression model on system workload.

Apart from the time-based optimal software rejuvenation schedule, some other stochastic models have been proposed in the literature. Bao et al. [30,31], Bobbio et al. [32], Okamura et al. [33], Wang et al. [34], and Xie et al. [35] developed condition-based software rejuvenation schemes, where the system workload is measured to trigger software rejuvenation. Although system parameters and resource usage strongly affect the software aging, it was observed that the mechanism of aging for individual software-based system has to be clarified. In other words, since the aging-related bugs may not be related to system workload explicitly, it may be difficult to apply the workload-based software rejuvenation to tolerate transient failures completely. Pfening et al. [36] formulate a server-type software system with degradation as a queuing system by a Markov decision process. Garg et al. [37] take account of the presence of system failure caused by software aging and analyze the time-based optimal software rejuvenation schedule. This model is extended latter in the study by Okamura et al. [38–40] by introducing a different workload-based rejuvenation schedule and/or a more general arrival process of transactions. Recently, Zheng et al. [41] generalized the existing stochastic models [37–40] in terms of the rejuvenation policies and arrival stream of transactions. Van Moorsel and Wolter [42] focus on the system restart and derive the optimal restart policies to rejuvenate a software system over finite and infinite operational periods.

It should be worth mentioning that all the above-mentioned software rejuvenation models are formulated under the assumption that the software system operates in continuous time. However, in real-world examples, it is not always possible to monitor and control software systems in continuous time. For example, it is common to back up the system data in file systems at any periodic timing, such as the end of every business hour or weekend. In this sense, the discrete-time models seem to be more realistic to make a decision for software rejuvenation. Dohi et al. [43] and Iwamoto et al. [44] proposed a software availability model and a software cost model operating in a discrete time setting, which are similar to those operating in continuous time [11,12]. Also Iwamoto et al. [45] formulate the discrete-time semi-Markov model under a different criterion called *cost-effectiveness* which is a mixture of system availability and expected cost in the long run. Iwamoto et al. [46] reformulate the MRGP model in continuous time in [17] and derive the optimal periodic software rejuvenation policy in discrete time by maximizing the steady-state system availability. Unfortunately, the discrete-time models with periodic rejuvenation have not been fully studied yet, so the analysis of expected cost model and cost-effectiveness models also remains incomplete. In this chapter we overview a software rejuvenation scheduling problem and provide a comprehensive survey of the optimal periodic software rejuvenation policies in discrete time under four criteria of optimality: expected cost per unit time in the steady state, steady-state system availability, cost-effectiveness, and expected total discounted cost over an infinite time horizon. The last three criteria are original results of this chapter. We characterize the respective optimal software rejuvenation policies in discrete-time model and provide their associated statistical estimation algorithms from the complete sample of failure time data.

The remaining part of this chapter is organized as follows. In Section 3.2, we describe an MRGP model with periodic software rejuvenation in accordance with Iwamoto et al. [46]. In Section 3.3, we formulate the expected cost per unit time in the steady state and derive the optimal periodic software rejuvenation time minimizing it. We also provide a statistically nonparametric estimator of the optimal periodic software rejuvenation time as a function of system failure time. Sections 3.4 and 3.5 concern the steady-state system availability and cost-effectiveness, respectively. We derive the analytical solutions and nonparametric estimators of the corresponding optimal periodic software rejuvenation policies. In Section 3.6, we consider a somewhat different criterion called the expected total discounted cost over an infinite time horizon. Dohi et al. [13] and Danjou et al. [10] investigate effects on discounting over the expected cost for software rejuvenation models with aperiodic policy [9] and periodic policy [16], respectively. On the other hand, a software cost model with discounting has not been known in discrete time. We formulate the expected total discounted cost over an infinite time horizon and minimize it. Numerical illustrations are given in Section 3.7, where the Monte Carlo simulation is conducted to investigate asymptotic behavior of estimators. Section 3.8 concludes the chapter with some remarks.

3.2 Model Description

3.2.1 Notation

Consider a periodic software rejuvenation model developed by Iwamoto et al. [46]. First, we define the system states and notation as follows:

State 0: highly robust state
State 1: failure probable state
State 2: failure state
State 3: software rejuvenation state
State 4: software rejuvenation state without through failure probable state.

The notation used in this chapter is as follows:

Z: time interval from highly robust state to failure probable state (discrete random variable)

$F_0(n)$, $f_0(n)$, $\mu_0(>0)$: cumulative distribution function (c.d.f.), probability mass function (p.m.f.), and mean of Z, where $n = 0, 1, 2, \ldots$

X: failure time from failure probable state (discrete random variable)

$F_f(n)$, $f_f(n)$, $\mu_f(>0)$: c.d.f., p.m.f., and mean of X

n_0: trigger time of software rejuvenation (integer value)

$F_a(n)$, $f_a(n)$, $\mu_a(>0)$: c.d.f., p.m.f., and mean of recovery operation time from failure state

$F_c(n)$, $f_c(n)$, $\mu_c(>0)$: c.d.f., p.m.f., and mean of overhead incurred by software rejuvenation

$c_s(>0)$: recovery (corrective maintenance) cost per unit time

$c_p(>0)$: rejuvenation (preventive maintenance) cost per unit time.

3.2.2 Model Description

The system operation of the software system starts at time $n = 0$ in the highly robust state. For some reason, such that the total amount of memory leaking attains a critical threshold, the process makes a transition to the failure probable state after the time period Z elapses. Just after the state becomes the failure probable state, a system failure may occur with positive probability. If the system failure occurs in the failure probable state before triggering a software rejuvenation, then the corrective recovery operation starts immediately at that time, and is completed after the random time with mean μ_a elapses. Otherwise, the software rejuvenation is triggered as a preventive maintenance of the software system, where the trigger time of software rejuvenation n_0 is measured from the beginning of system operation. This assumption is different from the previous aperiodic model by Dohi et al. [43] and Iwamoto et al. [44,45], because the trigger time of software rejuvenation under aperiodic policies is measured from the time instant when the failure probable state is observed. Of course, this may not be plausible in many cases if the system states (highly robust state and failure probable state) cannot be identified. After completing the recovery operation or software rejuvenation, the system state becomes as good as new, and the software age is initiated at the beginning of the next highly robust state. In the above model, note that the cycle length of software rejuvenation is measured from the time instant just after the system enters State 1 from State 0.

Let us consider the time to trigger software rejuvenation to be a constant integer n_0. We call $n_0 (\geq 0)$ the *software rejuvenation schedule*. Define the time interval from the beginning of the system operation to the completion of the preventive or corrective maintenance as one cycle. Figure 3.1 depicts the configuration of our model with periodic rejuvenation. Since the underlying stochastic process is a discrete MRGP with four regeneration states [46], we provide the transition diagram in Figure 3.2, where the states denoted by

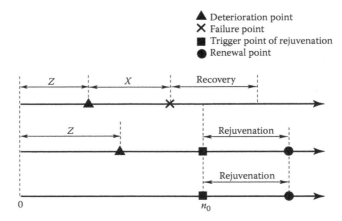

FIGURE 3.1
Configuration of periodic rejuvenation model in discrete time.

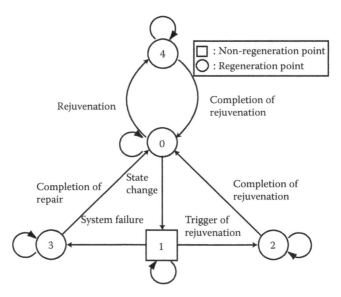

FIGURE 3.2
MRGP transition diagram.

circles (0, 2, 3, 4) and square (1) are regeneration points and non-regeneration point, respectively, in the MRGP. Strictly speaking, this stochastic process is not a common discrete-time semi-Markov process. However, since only one non-regeneration point is included in the transition diagram, it can be reduced to an equivalent discrete-time semi-Markov process by introducing the convolution of discrete probability distributions [46].

3.3 Expected Cost Analysis

3.3.1 Formulation

In this section, we consider the expected cost per unit time in the steady state. We make the following two assumptions:

Assumption (A-1): $c_s > c_p$,
Assumption (A-2): $\mu_a > \mu_c$.

The assumption (A-1) implies that the recovery cost per unit time is greater than the rejuvenation cost per unit time. On the other hand, in the assumption (A-2), the mean time of recovery operation is greater than the mean rejuvenation overhead. Under these plausible assumptions, the mean time length of one cycle and the expected total cost during one cycle are given by

$$T(n_0) = \mu_a (F_f * F_0)(n_0) + \mu_c \overline{(F_f * F_0)}(n_0) + \sum_{n=0}^{n_0-1} \overline{(F_f * F_0)}(n), \tag{3.1}$$

$$V(n_0) = c_s \mu_a (F_f * F_0)(n_0) + c_p \mu_c \overline{(F_f * F_0)}(n_0), \tag{3.2}$$

respectively, where $(F_f * F_0)(n)$ is the discrete Stieltjes convolution of $F_f(n)$ and $F_0(n)$, and $\overline{(F_f * F_0)}(n) = 1 - (F_f * F_0)(n)$ in general. Hence, the expected cost per unit time in the steady state is, from the renewal reward argument, given by

$$C(n_0) = \lim_{n \to \infty} \frac{E[\text{cost occured for }[0,n)]}{n} = \frac{V(n_0)}{T(n_0)}. \tag{3.3}$$

Then the problem is to seek the optimal software rejuvenation schedule n_0^* which minimizes $C(n_0)$.

Taking the difference of $C(n_0)$ with respect to n_0, we define the following function:

$$q_C(n_0) = \frac{T(n_0)T(n_0+1)[C(n_0+1)-C(n_0)]}{F_f * F_0(n_0)}$$

$$= (c_s\mu_a - c_p\mu_c)T(n_0)r(n_0+1) - V(n_0)$$

$$-(\mu_a - \mu_c)V(n_0)r(n_0+1), \qquad (3.4)$$

where $r(n) = (f_f * f_0)(n)/\overline{F_f * F_0}(n-1)$ is the discrete failure rate for the c.d.f. $(F_f * F_0)(n)$.

By checking the sign of Eq. (3.4), we can derive the following theorem to characterize the optimal periodic software rejuvenation policy.

Theorem 1

1. Suppose that the failure time distribution $(F_f * F_0)(n)$ is strictly IFR (increasing failure rate), that is, the failure rate $r(n)$ is strictly increasing in n, under the assumptions (A-1) and (A-2).

    i. If $q_C(0) < 0$ and $q_C(\infty) > 0$, then there exists (at least one, at most two) optimal software rejuvenation schedule n_0^* $(0 < n_0^* < \infty)$ satisfying $q_C(n_0^* - 1) < 0$ and $q_C(n_0^*) \geq 0$. The corresponding expected cost per unit time in the steady state $C(n_0^*)$ must satisfy

    $$K_C(n_0^*) \leq C(n_0^*) < K_C(n_0^* + 1), \qquad (3.5)$$

    where

    $$K_C(n) = \frac{(c_s\mu_a - c_p\mu_c)r(n)}{(\mu_a - \mu_c)r(n) + 1}. \qquad (3.6)$$

    ii. If $q_C(0) \geq 0$, then the optimal software rejuvenation schedule is $n_0^* = 0$, and the minimum expected cost per unit time in the steady state is given by

    $$C(0) = \frac{V(0)}{T(0)} = c_p. \qquad (3.7)$$

    iii. If $q_C(\infty) \leq 0$, then the minimum expected cost per unit time in the steady state is given by

    $$C(\infty) = \frac{V(\infty)}{T(\infty)} = \frac{c_s\mu_a}{\mu_a + \mu_f + \mu_0}. \qquad (3.8)$$

2. Suppose that the failure time distribution is DFR (decreasing failure rate), that is, $r(n)$ is decreasing in n, under the assumptions (A-1) and (A-2). Then, the optimal software rejuvenation schedule is $n_0^* = 0$ or $n_0^* \to \infty$.

Proof: Taking the difference of Eq. (3.4) yields

$$q_C(n_0+1) - q_C(n_0) = \{r(n_0+2) - r(n_0+1)\}$$
$$\times \{T(n_0+1)(c_s\mu_a - c_p\mu_c) - V(n_0+1)(\mu_a - \mu_c)\}. \quad (3.9)$$

From the assumptions (A-1) and (A-2), it follows from the reduction argument that $c_s + \mu_c(c_s - c_p)/(\mu_a - \mu_c) > C(n_0+1)$, so that the second term of the right-hand side of Eq. (3.9) must be strictly positive. If $(F_f * F_0)(n)$ is strictly IFR, then $q_C(n_0+1) > q_C(n_0)$. Furthermore, if $q_C(0) < 0$ and $q_C(\infty) > 0$, then the optimal software rejuvenation schedules n_0^* $(0 < n_0^* < \infty)$ satisfy $q_C(n_0^* - 1) < 0$ and $q_C(n_0^*) \geq 0$. These inequalities imply Eq. (3.5). If $q_C(0) \geq 0$ $(q_C(\infty) \leq 0)$, then $q_C(n_0)$ is always positive (negative) and $C(n_0)$ is an increasing (decreasing) function of n_0. On the other hand, if $(F_f * F_0)(n_0)$ is DFR, then $C(n_0)$ is a quasi-concave function of n_0, and the optimal software rejuvenation schedule is $n_0^* = 0$ or $n_0^* \to \infty$. Hence, the proof is completed.

3.3.2 Statistical Inference

In a fashion similar to the continuous case in [11–14], it can be shown that $(F_f * F_0)(n)$ is IFR (DFR) if and only if the function $\varnothing(p)$ is concave (convex) on $p \in [0,1]$, where

$$\varnothing(p) = \sum_{n=0}^{(F_f * F_0)^{-1}(p)} \frac{\overline{(F_f * F_0)}(n)}{\mu_f + \mu_0} \quad (3.10)$$

is called the discrete scaled total time on test (TTT) transform with

$$(F_f * F_0)^{-1}(p) = \min\{n : (F_f * F_0)(n) > p\} - 1, \quad (3.11)$$

if the inverse function exists. Then we have

$$\mu_f + \mu_0 = \sum_{n=0}^{\infty} \overline{(F_f * F_0)}(n). \quad (3.12)$$

From a few algebraic manipulations, we obtain the following theorem to interpret the underlying optimization problem $\min_{0 \leq n_0 < \infty} C(n_0)$ geometrically.

Theorem 2

Obtaining the optimal software rejuvenation schedule n_0^* by minimizing the expected cost per unit time in the steady state $C(n_0)$ is equivalent to obtaining p^* $(0 \leq p^* \leq 1)$ such as

$$\max_{0 \leq p \leq 1} \frac{\varnothing(p) + \alpha_C}{p + \beta_C}, \tag{3.13}$$

where

$$\alpha_C = \frac{\mu_c \mu_a (c_s - c_p)}{(\mu_f + \mu_0)(c_s \mu_a - c_p \mu_c)}, \quad \beta_C = \frac{c_p \mu_c}{c_s \mu_a - c_p \mu_c}. \tag{3.14}$$

Proof: From the definition in Eqs. (3.10) and (3.11), it is seen that

$$\min_{0 \leq n_0 < \infty} C(n_0) \Leftrightarrow \min_{0 \leq p \leq 1} \frac{\frac{(c_s \mu_a - c_p \mu_c)}{\mu_f + \mu_0} p + \frac{c_p \mu_c}{\mu_f + \mu_0}}{\varnothing(p) + \frac{\mu_c}{\mu_f + \mu_0} + \frac{\mu_a - \mu_c}{\mu_f + \mu_0} p}$$

$$\Leftrightarrow \max_{0 \leq p \leq 1} \frac{\varnothing(p) + \frac{\mu_c}{\mu_f + \mu_0} + \frac{\mu_a - \mu_c}{\mu_f + \mu_0} p}{\frac{(c_s \mu_a - c_p \mu_c)}{\mu_f + \mu_0} p + \frac{c_p \mu_c}{\mu_f + \mu_0}}$$

$$\Leftrightarrow \max_{0 \leq p \leq 1} \frac{\varnothing(p) + \frac{(c_s - c_p)\mu_a \mu_c}{(c_s \mu_a - c_p \mu_c)(\mu_f + \mu_0)}}{p + \frac{c_p \mu_c}{c_s \mu_a - c_p \mu_c}} = \frac{\varnothing(p) + \alpha_C}{p + \beta_C}. \tag{3.15}$$

Hence, the proof is completed.

From Theorem 2, it is seen that the optimal software rejuvenation schedule $n_0^* = (F_f * F_0)^{-1}(p^*)$ is determined by calculating the optimal point p^* $(0 \leq p^* \leq 1)$ maximizing the tangent slope from the point $(-\beta_C, -\alpha_C) \in (-\infty, 0) \times (-\infty, 0)$ to the curve $(p, \varnothing(p)) \in [0, 1] \times [0, 1]$. This graphical interpretation will be useful to estimate the optimal software rejuvenation schedule from the failure time data.

Next, suppose that the optimal software rejuvenation schedule has to be estimated from k ordered complete observations: $0 = x_0 \leq x_1 \leq x_2 \cdots \leq x_k$ of the failure times from a discrete c.d.f. $(F_f * F_0)(n)$, which is unknown. Then, the empirical distribution for this sample is given by

$$(F_f * F_0)_k(n) = \begin{cases} i/k, & \text{for } x_i \leq n < x_{i+1}, \\ 1, & \text{for } x_k \leq n. \end{cases} \quad (3.16)$$

The scaled TTT statistics based on this sample is defined by

$$\varnothing_{ik} = T_i / T_k, \quad i = 0, 1, 2, \ldots, k, \quad (3.17)$$

where

$$T_i = \sum_{j=1}^{i}(k-j+1)(x_j - x_{j-1}), \quad i = 1, 2, \ldots, k; \quad T_0 = 0. \quad (3.18)$$

The following theorem gives a statistically nonparametric estimation algorithm for the optimal software rejuvenation schedule.

Theorem 3

Suppose that the optimal software rejuvenation schedule has to be estimated from k ordered complete sample $0 = x_0 \leq x_1 \leq x_2 \cdots \leq x_k$ of the failure times from a discrete c.d.f. $(F_f * F_0)(n)$, which is unknown. Then, a nonparametric estimator of the optimal software rejuvenation schedule $\hat{n}_0^*$ which minimizes $C(n_0)$ is given by x_{j^*}, where

$$j^* = \left\{ j \mid \max_{0 \leq j \leq n} \frac{\varnothing_{jk} + \alpha_{Ck}}{j/k + \beta_C} \right\}, \quad (3.19)$$

with

$$\alpha_{Ck} = \frac{\alpha_C(\mu_f + \mu_0)}{\mu_{fk}}, \quad \mu_{fk} = \sum_{i=0}^{k} \frac{x_i}{k}. \quad (3.20)$$

In fact, since the empirical distribution is strongly consistent, that is, $(F_f * F_0)_k(n) \to (F_f * F_0)(n)$ as $k \to \infty$, the resulting estimator of the optimal software rejuvenation schedule is expected to asymptotically converge to the real (but unknown) optimal solution n_0^* under (A-1) and (A-2). In the subsequent sections, we apply the similar technique to the other optimality criteria.

3.4 Availability Analysis

3.4.1 Formulation

Next, we analyze the steady-state system availability. The mean operative time during one cycle is given by

$$S(n_0) = \sum_{n=0}^{n_0-1} (F_f * F_0)(n). \tag{3.21}$$

The system availability in the steady state for our periodic model is given by

$$A(n_0) = \frac{S(n_0)}{T(n_0)} \tag{3.22}$$

from the renewal reward argument. Then the problem is to seek the optimal software rejuvenation schedule n_0^* which maximizes $A(n_0)$.

Taking the difference of $A(n_0)$ with respect to n_0, we define the following function:

$$\begin{aligned} q_A(n_0) &= \frac{T(n_0)T(n_0+1)[A(n_0+1)-A(n_0)]}{(F_f * F_0)(n)} \\ &= T(n_0) - S(n_0) + (\mu_c - \mu_a)S(n_0)r(n_0+1). \end{aligned} \tag{3.23}$$

The optimal software rejuvenation schedule which maximizes the steady-state system availability is given in the following theorem:

Theorem 4

1. Suppose that the failure time distribution $(F_f * F_0)(n)$ is strictly IFR under the assumption (A-2).
   i. If $q_A(\infty) < 0$, then there exists (at least one, at most two) optimal software rejuvenation schedule n_0^* ($0 < n_0^* < \infty$) satisfying $q_A(n_0^*-1) > 0$ and $q_A(n_0^*) \leq 0$. The corresponding maximum steady-state system availability $A(n_0^*)$ must satisfy

$$K_A(n_0^*+1) \leq A(n_0^*) < K_A(n_0^*), \tag{3.24}$$

where

$$K_A(n) = \frac{1}{1+(\mu_a - \mu_c)r(n)}. \tag{3.25}$$

ii. If $q_A(\infty) \geq 0$, then the optimal software rejuvenation schedule is $n_0^* \to \infty$ and the maximum steady-state system availability is given by

$$A(\infty) = \frac{S(\infty)}{T(\infty)} = \frac{\mu_0 + \mu_f}{\mu_0 + \mu_f + \mu_a}. \tag{3.26}$$

2. Suppose that the failure time distribution is DFR under the assumption (A-2). Then, the steady-state system availability $A(n_0)$ is a quasi-convex function of n_0, and the optimal software rejuvenation schedule is $n_0^* \to \infty$.

3.4.2 Statistical Inference

Similar to the expected cost case, we obtain the following theorem by using the discrete scaled TTT transform:

Theorem 5

Obtaining the optimal software rejuvenation schedule n_0^* by maximizing the steady-state system availability $A(n_0)$ is equivalent to obtaining p^* $(0 \leq p^* \leq 1)$ such as

$$\max_{0 \leq p \leq 1} \frac{\varnothing(p)}{p + \beta_A}, \tag{3.27}$$

where

$$\beta_A = \frac{\mu_c}{\mu_a - \mu_c}. \tag{3.28}$$

By calculating the optimal point p^* $(0 \leq p^* \leq 1)$, by maximizing the tangent slope from the point $(-\beta_A, 0)$ to the curve $(p, \varnothing(p))$, we can determine the optimal software rejuvenation schedule n_0^* which maximizes the steady-state system availability on the graph.

Theorem 6

A nonparametric estimate of the optimal software rejuvenation schedule $\hat{n}_0^*$ which maximizes $A(n_0)$ is given by x_{j^*}, where

$$j^* = \left\{ j \mid \max_{0 \leq j \leq n} \frac{\varnothing_{jk}}{j/k + \beta_A} \right\}. \tag{3.29}$$

Note that β_A is independent of k.

3.5 Cost-Effectiveness Analysis

3.5.1 Formulation

Cost-effectiveness is defined by combining the expected cost per unit time in the steady state and the steady-state system availability. For analysis of cost-effectiveness, we make the following assumption:

Assumption (A-3): $c_s \mu_a > c_p \mu_c$.

We define the cost-effectiveness in the steady state as

$$E(n_0) = \lim_{n \to \infty} \frac{E[\text{operating time on } (0, n)]}{E[\text{cost incurred on } (0, n)]} = \frac{S(n_0)}{V(n_0)}. \tag{3.30}$$

The software rejuvenation schedule n_0^* which maximizes $E(n_0)$ becomes the optimal solution by taking account of both cost component and reliability component simultaneously.

Taking the difference of $E(n_0)$ with respect to n_0, we define the following function:

$$q_E(n_0) = \frac{V(n_0)V(n_0+1)[E(n_0+1)-E(n_0)]}{(F_f * F_0)(n_0)}$$

$$= V(n_0) - (c_s \mu_a - c_p \mu_c) S(n_0) r(n_0+1). \tag{3.31}$$

The optimal software rejuvenation schedule which maximizes the cost-effectiveness is given in the following theorem:

Theorem 7

1. Suppose that the failure time distribution $(F_f * F_0)(n)$ is strictly IFR under the assumption (A-3).
   i. If $q_E(\infty) < 0$, then there exists (at least one, at most two) optimal software rejuvenation schedule n_0^* ($0 < n_0^* < \infty$) satisfying $q_E(n_0^* - 1) > 0$ and $q_E(n_0^*) \leq 0$. The corresponding maximum cost-effectiveness $E(n_0^*)$ is given by

$$K_E(n_0^* + 1) \leq E(n_0^*) < K_E(n_0^*), \tag{3.32}$$

   where

$$K_E(n) = \frac{1}{(c_s \mu_a - c_p \mu_c) r(n)}. \tag{3.33}$$

ii. If $q_E(\infty) \geq 0$, then the optimal software rejuvenation schedule is $n_0^* \to \infty$, and the maximum cost-effectiveness is given by

$$E(\infty) = \frac{S(\infty)}{T(\infty)} = \frac{\mu_f + \mu_0}{c_s \mu_a}. \qquad (3.34)$$

2. Suppose that the failure time distribution is DFR under the assumption (A-3). Then, the cost-effectiveness $E(n_0)$ is a quasi-convex function of n_0, and the optimal software rejuvenation schedule is $n_0^* \to \infty$.

3.5.2 Statistical Inference

From the similarity to the results in Theorems 2 and 5, we obtain the following theorem by using the scaled TTT transform and the scaled TTT statistics:

Theorem 8

Obtaining the optimal software rejuvenation schedule n_0^* maximizing the cost-effectiveness $E(n_0)$ is equivalent to obtaining p^* $(0 \leq p^* \leq 1)$ such as

$$\max_{0 \leq p \leq 1} \frac{\varnothing(p)}{p + \beta_E}, \qquad (3.35)$$

where

$$\beta_E = \frac{c_p \mu_c}{c_s \mu_a - c_p \mu_c}. \qquad (3.36)$$

Theorem 9

A nonparametric estimate of the optimal software rejuvenation schedule $\hat{n}_0^*$ which maximizes $E(n_0)$ is given by x_{j^*}, where

$$j^* = \left\{ j \mid \max_{0 \leq j \leq n} \frac{\varnothing_{jk}}{j/k + \beta_E} \right\}. \qquad (3.37)$$

3.6 Expected Total Discounted Cost Analysis

3.6.1 Formulation

As the fourth optimality criterion, we introduce the expected total discounted cost over an infinite time horizon to take account of the effect of present value of cost function. Let $\gamma \in (0, 1)$ be a discount rate which denotes an interest rate in discrete time. To analyze the expected total discounted cost, we make the following assumptions:

Assumption (A-4): $\sum_{y=0}^{\infty} \gamma^y f_a(y) > \sum_{s=0}^{\infty} \gamma^s f_c(s)$,

Assumption (A-5): $c_s \sum_{y=0}^{\infty} \sum_{k=0}^{y} \gamma^k f_a(y) > c_p \sum_{s=0}^{\infty} \sum_{k=0}^{s} \gamma^k f_c(s)$.

The assumption (A-4) implies that the probability generating function of recovery operation time is greater than that of software rejuvenation overhead. On the other hand, the assumption (A-5) is somewhat technical but needed to show the existence of the optimal policy.

The discounted unit cost for one cycle and the expected total discounted cost for one cycle are given by

$$\delta(n_0) = \sum_{y=0}^{\infty} \sum_{x=0}^{n_0} \gamma^{x+y} (f_f * f_0)(x) f_a(y) + \sum_{s=0}^{\infty} \sum_{x=n_0+1}^{\infty} \gamma^{n_0+s} (f_f * f_0)(x) f_c(s),$$
(3.38)

$$V_\gamma(n_0) = c_s \sum_{y=0}^{\infty} \sum_{x=0}^{n_0} \sum_{k=1}^{y} \gamma^{k+x} (f_f * f_0)(x) f_a(y)$$

$$+ c_p \sum_{s=0}^{\infty} \sum_{x=n_0+1}^{\infty} \sum_{k=1}^{s} \gamma^{k+n_0} (f_f * f_0)(x) f_c(s),$$
(3.39)

respectively. Hence, the expected total discounted cost over an infinite time horizon is given by

$$TC(n_0) = \sum_{k=0}^{\infty} V_\gamma(n_0) \delta(n_0)^k = \frac{V_\gamma(n_0)}{\overline{\delta}(n_0)}.$$
(3.40)

Taking the difference of $TC(n_0)$ with respect to n_0, we define the following function:

$$q_\gamma(n_0) = \frac{\overline{\delta}(n_0) V_\gamma(n_0+1) - \overline{\delta}(n_0+1) V_\gamma(n_0)}{\gamma^{n_0+1} \overline{(F_f * F_0)}(n_0)}.$$
(3.41)

From Eq. (3.41), we can get the following theorems:

Theorem 10

1. Suppose that the failure time distribution $(F_f * F_0)(n)$ is strictly IFR under the assumptions (A-4) and (A-5).

   i. If $q_\gamma(0) < 0$ and $q_\gamma(\infty) > 0$, then there exists (at least one, at most two) optimal software rejuvenation schedule n_0^* ($0 < n_0^* < \infty$) satisfying $q_\gamma(n_0^* - 1) < 0$ and $q_\gamma(n_0^*) \geq 0$. The corresponding minimum expected total discounted cost over an infinite time horizon $TC(n_0^*)$ is given by

   $$K_\gamma(n_0^* + 1) \leq TC(n_0^*) < K_\gamma(n_0^*), \qquad (3.42)$$

   where

   $$K_\gamma(n) = \frac{c_s \sum_{y=0}^{\infty}\sum_{k=1}^{y}\gamma^k f_a(y) r(n) - c_p \sum_{s=0}^{\infty}\sum_{k=1}^{s}\gamma^k f_c(s)\left\{\dfrac{1}{\gamma} - 1 + r(n)\right\}}{\sum_{s=0}^{\infty}\gamma^s f_c(s) - \sum_{y=0}^{\infty}\gamma^y f_a(y) + \left(\dfrac{1}{\gamma} - 1\right)\sum_{s=0}^{\infty}\gamma^s f_c(s)}. \qquad (3.43)$$

   ii. If $q_\gamma(0) \geq 0$, then the optimal software rejuvenation schedule is $n_0^* = 0$ and the minimum expected total discounted cost over an infinite time horizon is given by

   $$TC(0) = \frac{V_\gamma(0)}{\delta(0)} = \frac{c_p \sum_{s=0}^{\infty}\sum_{k=1}^{s}\gamma^k f_c(s)}{1 - \sum_{s=0}^{\infty}\gamma^s f_c(s)}. \qquad (3.44)$$

   iii. If $q_\gamma(\infty) \leq 0$, then the optimal software rejuvenation schedule is $n_0^* \to \infty$ and the minimum expected total discounted cost over an infinite time horizon is given by

   $$TC(\infty) = \frac{V_\gamma(\infty)}{\delta(\infty)} = \frac{c_s \sum_{y=0}^{\infty}\sum_{x=0}^{n_0}\sum_{k=1}^{y}\gamma^{k+x}(f_f * f_0)(x) f_a(y)}{1 - \sum_{y=0}^{\infty}\sum_{x=0}^{\infty}\gamma^{x+y}(f_f * f_0)(x) f_a(y)}. \qquad (3.45)$$

2. Suppose that the failure time distribution is DFR under the assumptions (A-4) and (A-5). Then, the expected total discounted cost over an infinite time horizon $TC(n_0)$ is a quasi-concave function of n_0, and the optimal software rejuvenation schedule is $n_0^* \to \infty$.

3.6.2 Statistical Inference

For the expected total discounted cost, we define the following discrete modified scaled TTT transform:

$$\varnothing_\gamma(p) = \frac{1}{\tau_\gamma} \sum_{n=0}^{G^{-1}(p)} \gamma^x \overline{(F_f * F_0)}(s), \quad (3.46)$$

where

$$G^{-1}(p) = \min\left\{n_0 : 1 - \gamma^{n_0} \overline{(F_f * F_0)}(n_0) > p\right\} - 1, \quad (3.47)$$

if the inverse function exists. Then it holds that

$$\tau_\gamma = \sum_{x=0}^{\infty} \gamma^x \overline{(F_f * F_0)}(x). \quad (3.48)$$

Theorem 11

Obtaining the optimal software rejuvenation schedule n_0^* by minimizing the expected total discounted cost over an infinite time horizon $TC(n_0)$ is equivalent to obtaining $p^* \left(0 \le p^* \le 1\right)$ such as

$$\max_{0 \le p \le 1} \frac{\varnothing_\gamma(p) + \alpha_\gamma}{p + \beta_\gamma}, \quad (3.49)$$

where

$$\alpha_\gamma = \frac{a(1-d) - b(1-c)}{(1-\gamma)(ad - bc)\tau_y}, \quad \beta_\gamma = \frac{a}{ad - bc} - 1, \quad (3.50)$$

$$a = c_s \sum_{y=0}^{\infty} \sum_{k=1}^{y} \gamma^k f_a(y), \quad b = c_p \sum_{s=0}^{\infty} \sum_{k=1}^{s} \gamma^k f_c(s), \quad (3.51)$$

$$c = \sum_{y=0}^{\infty} \gamma^y f_a(y), \quad d = \sum_{s=0}^{\infty} \gamma^s f_c(s). \quad (3.52)$$

From this result, we can determine the optimal software rejuvenation schedule n_0^* which minimizes the expected total discounted cost over an infinite time horizon by calculating the optimal point $p^* \left(0 \le p^* \le 1\right)$ maximizing the tangent slope from the point $\left(-\beta_\gamma, -\alpha_\gamma\right) \in (-\infty, 0) \times (-\infty, 0)$ to the curve $\left(p, \varnothing(p)\right) \in [0, 1] \times [0, 1]$.

Next, we consider the statistically nonparametric estimation algorithm from k ordered complete samples. The modified empirical distribution for this sample is given by

$$(F_f * F_0)_{\gamma k}(n) = \begin{cases} 1 - \gamma^{x_i}(1 - i/k), & \text{for } x_i \leq n < x_{i+1}, \\ 1, & \text{for } x_k \leq n. \end{cases} \quad (3.53)$$

Then the numerical counterpart of the discrete modified TTT transform in Eq. (3.46) based on this sample, is defined by

$$\varnothing_{\gamma ik} = T_{\gamma i}/T_{\gamma k}, \quad i = 0, 1, 2, \ldots, k, \quad (3.54)$$

where

$$T_{\gamma i} = \sum_{j=1}^{i} \left(\frac{\gamma}{1-\gamma} \right)(k - j + 1)(x_j - x_{j-1})(1 - \gamma^{x_j}), \quad i = 1, 2, \ldots, k; \; T_{\gamma 0} = 0. \quad (3.55)$$

From Eqs. (3.54) and (3.55), we derive the following theorem:

Theorem 12

Suppose that the optimal software rejuvenation schedule has to be estimated from k ordered complete sample $0 = x_0 \leq x_1 \leq x_2 \cdots \leq x_k$ of the failure times from a discrete c.d.f. $(F_f * F_0)(n)$, which is unknown. Then, a nonparametric estimator of the optimal software rejuvenation schedule $\hat{n}_0^*$ which minimizes $TC(n_0)$ is given by x_{j^*}, where

$$j^* = \left\{ j \mid \max_{0 \leq j \leq k} \frac{\varnothing_{\gamma jk} + \alpha_{\gamma k}}{1 - \gamma^{x_j}(1 - j/k) + \beta_\gamma} \right\}, \quad (3.56)$$

where

$$\alpha_{\gamma k} = \frac{\alpha_\gamma \tau_\gamma}{\tau_{\gamma k}}, \quad \tau_{\gamma k} = \sum_{i=0}^{k} \gamma^{x_i}(1 - i/k)(x_{i+1} - x_i). \quad (3.57)$$

3.7 Numerical Examples

In this section, we present some examples to determine numerically the optimal software rejuvenation schedules. Suppose that the time to failure distribution of X is given by the negative binomial distribution with p.m.f.:

$$f_f(x) = \binom{x-1}{r-1} q^r (1-q)^{x-r}, \tag{3.58}$$

and the degradation time Z obeys the geometrical distribution with p.m.f.:

$$f_0(x) = \xi(1-\xi)^x, \tag{3.59}$$

where $q \in (0, 1)$, $\xi \in (0, 1)$ and $r = 1, 2, \ldots$ is the natural number. In the remaining part of this section, we assume that $(r, q) = (10, 0.3)$, $\xi = 0.3$, $c_s = 5.0 \times 10[\$/\text{day}]$, $c_p = 4.0 \times 10[\$/\text{day}]$, $\mu_a = 5.0$, and $\mu_c = 2.0$.

3.7.1 Case of Expected Cost

The left-hand side of Figure 3.3 illustrates the determination of the optimal software rejuvenation schedule on the two-dimensional graph under the expected cost per unit time. Since $p^* = 0.0742365$ has the maximum slope from $(-\beta_C, -\alpha_C) = (-0.470558, -0.164926)$ in Figure 3.3, the optimal software rejuvenation schedule is given by $n_0^* = (F_f * F_0)^{-1} (0.0742365) = 24$. Then, the corresponding expected cost per unit time in the steady state is $C(24) = 3.69544$.

The right-hand side of Figure 3.3 shows the estimation result of the optimal software rejuvenation schedule under the expected cost per unit time, where the failure time data are generated from the negative binomial distribution in Eq. (3.58) and the geometric distribution in Eq. (3.59). For 200 simulation data, the estimates of the optimal rejuvenation schedule and the minimum expected cost are given by $\hat{n}_0^* = x_8 = 22$ and $C(\hat{n}_0^*) = 3.58349$, respectively.

Figure 3.4 illustrates the asymptotic behavior of estimates of the optimal software rejuvenation schedule and its minimum expected cost per unit time. In this figure, estimates of n_0^* and $C(\hat{n}_0^*)$ are calculated in accordance with the estimation algorithm given in Theorem 3, where the horizontal lines denote the real optimal solutions. From Figure 3.4, it is seen that

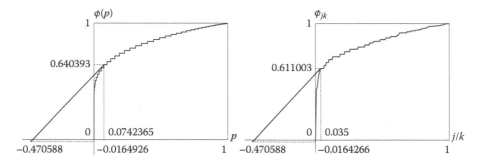

FIGURE 3.3
Determination and estimation of the optimal software rejuvenation schedule under expected cost per unit time.

Optimal Periodic Software Rejuvenation Policies

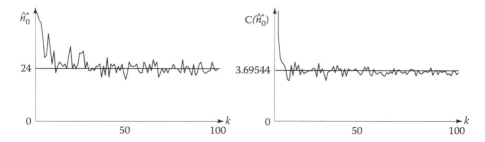

FIGURE 3.4
Asymptotic behavior of estimates of the optimal software rejuvenation schedule and its minimum expected cost per unit time.

the optimal software rejuvenation schedule and the expected cost per unit time can be estimated accurately from around $k = 25$. These observations enable us to use the nonparametric algorithm proposed here for precisely estimating the optimal software rejuvenation schedules and their associated expected costs per unit time, under the incomplete knowledge of the failure time distribution.

3.7.2 Case of System Availability

The left-hand side of Figure 3.5 presents the determination of the optimal software rejuvenation schedule on the two-dimensional graph under the availability criterion. Since $p^* = 0.125921$ has the maximum slope from $(-\beta_A, 0) = (-0.666667, 0)$ in Figure 3.5, the optimal software rejuvenation schedule is $n_0^* = \left(F_f * F_0\right)^{-1}(0.125921) = 26$. Then, the corresponding steady-state system availability is $A(26) = 0.91215$.

The right-hand side of Figure 3.5 illustrates the estimation result of the optimal software rejuvenation schedule, where the failure time data are generated from the negative binomial distribution in Eq. (3.58). For 200 pseudo

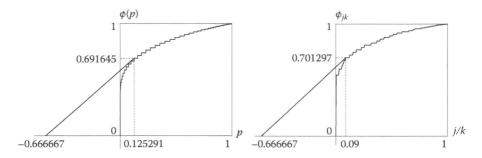

FIGURE 3.5
Determination and estimation of the optimal software rejuvenation schedule under steady-state system availability.

random numbers, the estimates of the optimal rejuvenation schedule and its maximum system availability are given by $\hat{n}_0^* = x_{19} = 26$ and $A(\hat{n}_0^*) = 0.918798$, respectively.

Figure 3.6 shows the asymptotic behavior of estimates of the optimal software rejuvenation schedule and its maximum system availability. In the figure, estimates of n_0^* and $A(\hat{n}_0^*)$ are calculated from the estimation algorithm given in Theorem 6, where the horizontal lines in the figure denote the real optimal solutions. From Figure 3.6, it is seen that the optimal software rejuvenation schedule and its associated system availability can be estimated accurately around $k = 20$, respectively.

3.7.3 Case of Cost-Effectiveness

The left-hand side of Figure 3.7 illustrates the determination of the optimal software rejuvenation schedule on the two-dimensional graph under the cost-effectiveness criterion. Since $p^* = 0.0742365$ has the maximum slope from $(-\beta_E, 0) = (-0.470588, 0)$ in Figure 3.7, the optimal software

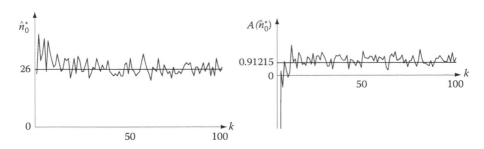

FIGURE 3.6
Asymptotic behavior of estimates of the optimal software rejuvenation schedule and its maximum system availability.

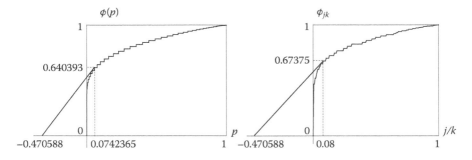

FIGURE 3.7
Determination and estimation of the optimal software rejuvenation schedule under cost-effectiveness.

rejuvenation schedule is $n_0^* = (F_f * F_0)^{-1}(0.0742365) = 24$, and the corresponding cost-effectiveness is $E(24) = 0.246606$.

The right-hand side of Figure 3.7 shows the estimation results of the optimal software rejuvenation schedule. For 200 simulation data, the estimates of the optimal rejuvenation schedule and the maximum cost-effectiveness are given by $\hat{n}_0^* = x_{17} = 25$ and $E(\hat{n}_0^*) = 0.264209$, respectively.

Figure 3.8 illustrates the asymptotic behavior of estimates of the optimal software rejuvenation schedule and its maximum cost-effectiveness. In the figure, the estimates of n_0^* and $E(\hat{n}_0^*)$ are calculated in accordance with the estimation algorithm given in Theorem 9. From Figure 3.8, it is seen that the optimal software rejuvenation schedule and the cost-effectiveness can be estimated accurately from around $k = 20$.

3.7.4 Case of Expected Total Discounted Cost

In the expected total discounted cost case, we assume that $(r, q = 12, 0.5)$, $\gamma = 0.5$, $c_s = 5.0 \times 10 [\$/\text{day}]$, $c_p = 3.0 \times 10 [\$/\text{day}]$, $\sum_{y=0}^{\infty} \gamma^y f_a(y) = 0.3$, $\sum_{s=0}^{\infty} \gamma^s f_c(s) = 0.8$, and $\gamma = 0.97$.

The left-hand side of Figure 3.9 shows the determination of the optimal software rejuvenation schedule on the two-dimensional graph in the case of expected total discounted cost. Since $p^* = 0.400808$ has the maximum slope from $(-\beta_\gamma, -\alpha_\gamma) = (-0.335878, -0.202587)$ in Figure 3.9, the optimal software rejuvenation schedule is $n_0^* = G^{-1}(0.400808) = 17$ in Eq. (3.49). Then, the corresponding expected total discounted cost over an infinite time horizon is given by $TC(17) = 258.939$.

The right-hand side of Figure 3.9 presents the estimation results of the optimal software rejuvenation schedule. For 200 simulation data, the estimates of the optimal rejuvenation schedule and its minimum expected total discounted cost are $\hat{n}_0^* = x_9 = 20$ and $TC(\hat{n}_0^*) = 270.522$, respectively.

Figure 3.10 illustrates the asymptotic behavior of estimates of the optimal software rejuvenation schedule and its minimum expected total discounted

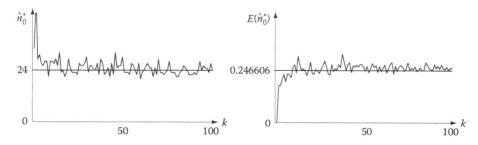

FIGURE 3.8
Asymptotic behavior of estimates of the optimal software rejuvenation schedule and its maximum cost-effectiveness.

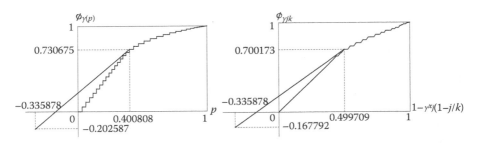

FIGURE 3.9
Determination and estimation of the optimal software rejuvenation schedule under expected total discounted cost.

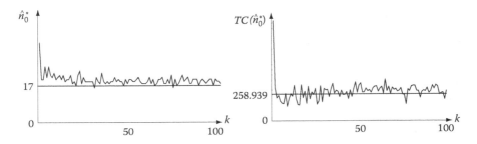

FIGURE 3.10
Asymptotic behavior of estimates of the optimal software rejuvenation schedule and its minimum expected total discounted cost.

cost. In the figure, the estimates of n_0^* and $TC(\hat{n}_0^*)$ are calculated using Theorem 12. From Figure 3.10, it is seen that the optimal software rejuvenation schedule and the expected total discounted cost can be estimated accurately in rather early phase, respectively.

3.8 Conclusions

In this chapter, we have summarized optimal periodic software rejuvenation policies in discrete time and their statistical inference approach under the incomplete knowledge of failure time distribution. The key idea is to apply the discrete scaled TTT transform and its numerical counterpart. In this context, the results for expected cost per unit time in the steady state were presented in [46], but the other results for the system availability, cost-effectiveness, and the expected total discounted cost have not been known yet.

In the future, we will consider the discrete models for NPI of software rejuvenation schedule in a similar approach to the references [24–26].

References

1. E. Adams, Optimizing preventive service of the software products, *IBM Journal of Research & Development*, vol. 28, no. 1, pp. 2–14, 1984.
2. A. Avritzer and E.J. Weyuker, Monitoring smoothly degrading systems for increased dependability, *Empirical Software Engineering*, vol. 2, no. 1, pp. 59–57, 1997.
3. V. Castelli, R.E. Harper, P. Heidelberger, S.W. Hunter, K.S. Trivedi, K.V. Vaidyanathan, and W.P. Zeggert, Proactive management of software aging, *IBM Journal of Research & Development*, vol. 45, no. 2, pp. 311–332, 2001.
4. M. Grottke, L. Lie, K.V. Vaidyanathan, and K.S. Trivedi, Analysis of software aging in a web server, *IEEE Transactions on Reliability*, vol. 55, no. 3, pp. 411–420, 2006.
5. A.T. Tai, L. Alkalai, and S.N. Chau, On-board preventive maintenance: a design-oriented analytic study for long-life applications, *Performance Evaluation*, vol. 35, no. 3, pp. 215–232, 1999.
6. W. Yurcik and D. Doss, Achieving fault-tolerant software with rejuvenation and reconfiguration, *IEEE Computer*, vol. 18, no. 4, pp. 48–52, 2001.
7. J. Gray and D.P. Siewiorek, High-availability computer systems, *IEEE Computer*, vol. 24, no. 9, pp. 39–48, 1991.
8. M. Grottke, and K.S. Trivedi, Fighting bugs: Remove, retry, replicate, and rejuvenate, *IEEE Computer*, vol. 40, pp. 107–109, 2007.
9. Y. Huang, C. Kintala, N. Kolettis, and N.D. Fulton, Software rejuvenation: analysis, module and applications, in *Proceedings of the 25th International Symposium on Fault Tolerant Computing (FTC-1995)*, pp. 381–390, IEEE CPS, 1995.
10. T. Danjou, T. Dohi, N. Kaio, and S. Osaki, Analysis of periodic software rejuvenation policies based on net present value approach, *International Journal of Reliability, Quality and Safety Engineering*, vol. 11, no. 4, pp. 313–327, 2004.
11. T. Dohi, K. Goseva-Popstojanova, and K.S. Trivedi, Analysis of software cost models with rejuvenation, in *Proceedings of the 5th IEEE International Symposium on High Assurance Systems Engineering (HASE-2000)*, pp. 25–34, IEEE CPS, 2000.
12. T. Dohi, K. Goseva-Popstojanova, and K.S. Trivedi, Statistical non-parametric algorithms to estimate the optimal software rejuvenation schedule, in *Proceedings of 2000 Pacific Rim International Symposium on Dependable Computing (PRDC-2000)*, pp. 77–84, IEEE CPS, 2000.
13. T. Dohi, T. Danjou, and H. Okamura, Optimal software rejuvenation policy with discounting, in *Proceedings of 2001 Pacific Rim International Symposium on Dependable Computing (PRDC-2001)*, pp. 87–94, IEEE CPS, 2001.
14. T. Dohi, K. Goseva-Popstojanova, and K.S. Trivedi, Estimating software rejuvenation schedule in high assurance systems, *The Computer Journal*, vol. 44, no. 6, pp. 473–485, 2001.

15. T. Dohi, H. Suzuki, and S. Osaki, Transient cost analysis of non-Markovian software systems with rejuvenation, *International Journal of Performability Engineering*, vol. 2, no. 3, pp. 233–243, 2006.
16. S. Garg, M. Telek, A. Puliafito, and K.S. Trivedi, Analysis of software rejuvenation using Markov regenerative stochastic Petri net, in *Proceedings of 6th International Symposium on Software Reliability Engineering (ISSRE-1995)*, pp. 24–27, IEEE CPS, 1995.
17. H. Suzuki, T. Dohi, K. Goseva-Popstojanova, and K.S. Trivedi, Analysis of multi-step failure models with periodic software rejuvenation, in *Advances in Stochastic Modelling* (eds., J.R. Artalejo and A. Krishnamoorthy), pp. 85–108, Notable Publications, Neshanic Station, NJ, 2002.
18. T. Dohi, H. Okamura, and K.S. Trivedi, Optimizing software rejuvenation policies under interval reliability criteria, in *Proceedings of the 9th IEEE International Conference on Autonomic and Trusted Computing (ATC-2012)*, pp. 478–485, IEEE CPS, 2012.
19. J. Zhao, W.B. Wang, G.R. Ning, K.S. Trivedi, R. Matias Jr., and K.Y. Cai, A comprehensive approach to optimal software rejuvenation, *Performance Evaluation*, vol. 70, no. 11, pp. 917–933, 2013.
20. K. Rinsaka and T. Dohi, Optimizing software rejuvenation schedule based on the kernel density estimation, *Quality Technology & Quantitative Management*, vol. 6, no. 1, pp. 55–65, 2009.
21. R.P.W. Duin, On the choice of smoothing parameters for Parzen estimators of probability density functions, *IEEE Transactions on Computers*, vol. C-25, no. 11, pp. 1175–1179, 1976.
22. E. Parzen, On the estimation of a probability density function and the mode, *Annals of Mathematical Statistics*, vol. 33, no. 3, pp. 1065–1076, 1962.
23. B.W. Silverman, *Density Estimation for Statistics and Data Analysis*, Chapman & Hall, London, 1986.
24. K. Rinsaka and T. Dohi, Non-parametric predictive inference of preventive rejuvenation schedule in operational software systems, in *Proceedings of the 18th International Symposium on Software Reliability Engineering (ISSRE-2007)*, pp. 247–256, IEEE CPS, 2007.
25. K. Rinsaka and T. Dohi, Toward high assurance software systems with adaptive fault management, *Software Quality Journal*, vol. 24, pp. 65–85, 2016.
26. K. Rinsaka and T. Dohi, An adaptive cost-based software rejuvenation scheme with nonparameteric predictive inference approach, *Journal of the Operations Research Society of Japan*, vol. 60, no. 4, pp. 461–478, 2017.
27. F.P.A. Coolen and K.J. Yan, Nonparametric predictive inference with right-censored data, *Journal of Statistical Planning and Inference*, vol. 126, no. 1, pp. 25–54, 2004.
28. P. Coolen-Schrijner and F.P.A. Coolen, Adaptive age replacement strategies based on nonparametric predictive inference, *Journal of the Operational Research Society*, vol. 55, pp. 1281–129, 2004.
29. K.V. Vaidyanathan and K.S. Trivedi, A comprehensive model for software rejuvenation, *IEEE Transactions on Dependable and Secure Computing*, vol. 2, no. 2, pp. 124–137, 2005.

30. Y. Bao, X. Sun, and K.S. Trivedi, Adaptive software rejuvenation: degradation model and rejuvenation scheme, in *Proceedings of 33rd Annual IEEE/IFIP International Conference on Dependable Systems and Networks (DSN-2003)*, pp. 241–248, IEEE CPS, 2003.
31. Y. Bao, X. Sun, and K.S. Trivedi, A workload-based analysis of software aging, and rejuvenation, *IEEE Transactions on Reliability*, vol. 54, no. 3, pp. 541–548, 2005.
32. A. Bobbio, M. Sereno, and C. Anglano, Fine grained software degradation models for optimal rejuvenation policies, *Performance Evaluation*, vol. 46, no. 1, pp. 45–62, 2001.
33. H. Okamura, H. Fujio, and T. Dohi, Fine-grained shock models to rejuvenate software systems, *IEICE Transactions on Information and Systems (D)*, vol. E86-D, no. 10, pp. 2165–2171, 2003.
34. D. Wang, W. Xie, and K.S. Trivedi, Performability analysis of clustered systems with rejuvenation under varying workload, *Performance Evaluation*, vol. 64, no. 3, pp. 247–265, 2007.
35. W. Xie, Y. Hong, and K.S. Trivedi, Analysis of a two-level software rejuvenation policy, *Reliability Engineering and System Safety*, vol. 87, no. 1, pp. 13–22, 2005.
36. S. Pfening, S. Garg, A. Puliafito, M. Telek, and K.S. Trivedi, Optimal rejuvenation for tolerating soft failure, *Performance Evaluation*, vol. 27/28, no. 4, pp. 491–506, 1996.
37. S. Garg, A. Puliafito, M. Telek, and K.S. Trivedi, Analysis of preventive maintenance in transactions based software systems, *IEEE Transactions on Computers*, vol. 47, pp. 96–107, 1998.
38. H. Okamura, S. Miyahara, T. Dohi, and S. Osaki, Performance evaluation of workload-based software rejuvenation schemes, *IEICE Transactions on Information and Systems (D)*, vol. E84-D, no. 10, pp. 1368–1375, 2001.
39. H. Okamura, S. Miyahara, and T. Dohi, Dependability analysis of a transaction-based multi-server system with rejuvenation, *IEICE Transactions on Fundamentals of Electronics, Communications and Computer Sciences (A)*, vol. E86-A, no. 8, pp. 2081–2090, 2003.
40. H. Okamura, S. Miyahara, and T. Dohi, Rejuvenating communication network system with burst arrival, *IEICE Transactions on Communications (B)*, vol. E88-B, no. 12, pp. 4498–4506, 2005.
41. J. Zheng, H. Okamura, L. Li, and T. Dohi, A comprehensive evaluation of software rejuvenation policies for transaction systems with Markovian arrivals, *IEEE Transactions on Reliability*, vol. 66, no. 4, pp.1157–1177, 2017.
42. A.P.A. van Moorsel and K. Wolter, Analysis of restart mechanisms in software systems, *IEEE Transactions on Software Engineering*, vol. 32, no. 8, pp. 547–558, 2006.
43. T. Dohi, K. Iwamoto, H. Okamura, and N. Kaio, Discrete availability models to rejuvenate a telecommunication billing application, *IEICE Transactions on Communications (B)*, vol. E86-B, no. 10, pp. 2931–2939, 2003.
44. K. Iwamoto, T. Dohi, H. Okamura, and N. Kaio, Discrete-time cost analysis for a telecommunication billing application with rejuvenation, *Computers & Mathematics with Applications*, vol. 51, no. 2, pp. 335–344, 2006.

45. K. Iwamoto, T. Dohi, and N. Kaio, Estimating discrete-time periodic software rejuvenation schedules under cost effectiveness criterion, *International Journal of Reliability, Quality and Safety Engineering*, vol. 13, no. 6, pp. 565–580, 2006.
46. K. Iwamoto, T. Dohi, and N. Kaio, Estimating periodic software rejuvenation schedule in discrete operational circumstance, *IEICE Transactions on Information and Systems (D)*, vol. E91-D, no. 1, pp. 23–31, 2008.

4

Potential Applications of Multivariate Analysis for Modeling the Reliability of Repairable Systems—Examples Tested

Miguel Angel Navas, Carlos Sancho, and Jose Carpio
Spanish National Distance Education University

CONTENTS

4.1 Introduction: Background and Driving Forces 109
4.2 Study of the Reliability of the System with Methods of IEC Standards 110
4.3 The Complex Nature of Failures 117
4.4 Three-Dimensional Graphical Representation of $z(t)$ of the Repairable System 120
4.5 Potential Multivariate Applications for the Reliability Analysis of Repairable Systems 121
4.6 Correlations Analysis 122
4.7 Principal Component Analysis 129
4.8 Factor Analysis 129
4.9 Cluster Analysis 130
4.10 Canonical Correlation Analysis 132
4.11 Correspondence Analysis 133
4.12 Regression Analysis 133
4.13 Discriminant Analysis 136
4.14 Conclusions 137
References 138

4.1 Introduction: Background and Driving Forces

In the last few decades, the scientific community has developed various statistical methodologies for analysis and modeling of the reliability of repairable systems. The most widely used and accepted methods have been collected by the International Electrotechnical Commission (IEC) in its TC56 technical committee "Dependability," which has published standards for the

European region, taking into account military manuals (MIL-HDBK) issued by the Department of Defense of the United States of America.

The applicable statistical methods and analysis procedures are included in IEC 60300-3-5 (2001). The failure intensity $z(t)$ of a repairable item can be estimated using the successive time between failures (TBF), by means of a stochastic process (SP). If TBF shows no trend and is distributed exponentially, $z(t)$ is constant, and in this case it can be modeled by a homogenous Poisson process (HPP). In cases where there is a trend in $z(t)$, a nonhomogeneous Poisson process (NHPP) can be applied, through a power-law process (PLP).

The IEC 60605-6 developed procedures to determine whether or not a trend in $z(t)$ exists using the U statistic, both for a repairable system and for identical items of repairable systems. If there is no trend in $z(t)$, the point estimation of parameters and confidence intervals is performed with the standard IEC 60605-4. If there is a trend in $z(t)$, the parameters of the PLP model are estimated using the IEC 61710 standard.

Here, the methodology proposed by the IEC is applied to electric traction systems in three series of trains and one series of escalators, from which operating data were available for a period of more than 10 years. Tests of the electric traction systems of the 5000-4th series of trains are presented, and a complementary multivariate analysis is then proposed to characterize the $z(t)$ obtained and analyze the influence of recurrent failures.

4.2 Study of the Reliability of the System with Methods of IEC Standards

As already mentioned, in the statistical methods developed in the IEC standards, the $z(t)$ of the repairable systems can be estimated using an SP. If there is no trend, IEC standards assume that the failures are distributed exponentially and that the number of failures per unit time can be modeled by HPP "perfect repair" (same as new), in which case $z(t)$ is constant. For cases where there is a trend in $z(t)$, NHPP "minimal repair" (same as old) is applied, with modeling using PLP. This method that is most widely used in the industry and has also been adopted by the IEC was developed by Crow (1975).

The IEC standards do not support renewal process (RP) modeling of "perfect repair" when there is no trend in $z(t)$ and the TBF is not distributed exponentially. The nonexistence of a contrasted trend using the statistic U does not guarantee that the TBF will be distributed exponentially. Nor do they envisage modeling with an alternative NHPP for PLP, for which numerous contributions have been published for the improvement of PLP (Attardi and Pulcini, 2005; Bettini et al., 2007).

The procedures to be used in a railway company should preferably be standardized by independent international bodies, in order to be able to demonstrate objectively the safety of their maintenance processes; see

EN 50126-1 (1999) for the specification and demonstration of reliability, availability, maintainability, and security (RAMS).

The recommendations of IEC 60300-3-1 (2003) for the selection of analysis techniques and IEC 61703 (2001) for the use of mathematical expressions are also used here. The investigation was carried out in four underground railway repairable systems:

a. The electric traction system of 36 trains of the 5000-4th series
b. The electric traction system of 88 trains of the 2000-B series
c. The electric traction system of 23 trains of the 8000 series
d. The system of 40 escalators of the TNE model

The traction systems of the trains are composed of repairable items. In the case of failure of an item, for example, a traction motor, the item is removed for repair and once repaired is reassembled on a train. The traction systems are predominantly electric and electronic.

The repairs performed on these systems usually involve the exchange of a damaged component with another, with the assumption that this repair will restore the system to its initial operating state, that is, a "perfect repair." RP models (including the HPP) are most suitable for a priori use to model reliability (Figure 4.1).

The results of tests on the electric traction systems of the 5000-4th series trains are presented in this chapter (note that the results and conclusions of the same tests carried out on the other three repairable systems were very similar).

The previous studies on the reliability of repairable railway systems using alternative models to SPs include the work of Anderson and Peters (1993) in locomotives; Yongqin and Xishi (1996) in an automatic train protection (ATP) system; Bozzo et al. (2003), Sagareli (2004), and Chen et al. (2007) in AC

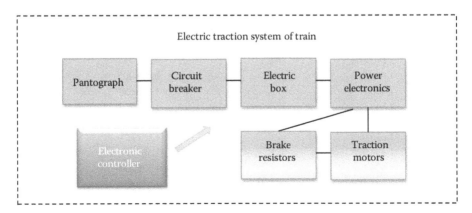

FIGURE 4.1
Repair blocks of an electric traction system.

traction systems. For studies using SP models, see the work of Panja and Ray (2007) on signaling and Luo et al. (2010) on brake control.

For electric traction systems, distance in kilometers (km) can be used as a variable instead of time (t) for reliability studies, since the full operation of traction systems depends on the train having traveled some distance, not on the passage of time itself. For example, Anderson and Peters (1993) used kilometers in their study, as traction systems are subjected to wear and tear by the mileage covered.

In the traction systems of 36 trains of the 5000-4th series studied, the records of failures corresponded to the period 1993–2008, that is, 16 years of commercial use. The mileage accumulated for each train exceeded 1,500,000 km. The total number of failures recorded during the study period was 3,112 (Table 4.1).

There is an important and detailed database existing in the railway company, which ensures that the results obtained in the statistical analyses have a high degree of integrity.

In this study, first we attempt to model the reliability of the traction systems of the trains, by means of the estimation of parameters for multiple items, using a single estimation to explain the behavior of the failures of the totality of each system. If this approach fails, the analysis is performed item by item, that is, for each train independently, in order to model the reliability obtained in each train in concrete terms, as recommended by IEC standards. This procedure was also described by Rigdon and Basu (2000).

TABLE 4.1

Summary of Failure Data of Traction Items in 5000-4th Trains

Train	Total Kilometers	Total Failures	Train	Total Kilometers	Total Failures
1	1,558,028	95	19	1,573,678	69
2	1,596,175	97	20	1,587,731	67
3	1,603,409	121	21	1,651,900	75
4	1,633,340	77	22	1,601,105	84
5	1,610,401	107	23	1,594,275	88
6	1,682,072	103	24	1,547,130	60
7	1,672,608	84	25	1,596,740	78
8	1,630,363	99	26	1,491,402	90
9	1,593,031	73	27	1,615,831	100
10	1,643,983	126	28	1,621,377	62
11	1,637,886	118	29	1,574,576	72
12	1,626,085	73	30	1,551,079	100
13	1,653,920	111	31	1,557,868	68
14	1,634,162	67	32	1,529,434	66
15	1,631,779	92	33	1,572,606	79
16	1,647,438	96	34	1,568,278	80
17	1,682,277	64	35	1,574,030	85
18	1,589,800	79	36	1,622,617	107

Potential Applications of Multivariate Analysis

The U-test for multiple items is then applied under IEC 60605-6 (2007) in Section 7.3. With r_i the total number of failures to consider from the ith item, T_i^* the total time (or km) of the test for the ith item, T_{ij} the time (or km) accumulated at the jth failure of the ith item, and k the total number of items:

$$U = \frac{\sum_{i=1}^{k}\sum_{j=1}^{r_i} T_{ij} - 0,5(r_1 T_1^* + r_2 T_2^* + \cdots r_k T_k^*)}{\sqrt{\frac{1}{12}(r_1 T_1^{*2} + r_2 T_2^{*2} + \cdots r_k T_k^{*2})}}. \tag{4.1}$$

The U statistic (Laplace test) is distributed approximately, according to a distribution typified by average 0 and deviation 1. The U statistic can be used to test whether there is evidence of positive or negative growth of reliability, independent of its pattern of growth.

A bilateral test for positive or negative growth with significance level α has critical values $u_{1-\alpha/2}$ and $-u_{1-\alpha/2}$, where $u_{1-\alpha/2}$ is the $(1-\alpha/2)100\%$ percentile of the typical normal distribution. If $-u_{1-\alpha/2} < U < u_{1-\alpha/2}$, then there is no evidence of positive or negative growth of the reliability to a significance level α. In this case, the hypothesis of an exponential distribution of times between successive failures of the HPP is accepted with significance level α. The critical values $u_{1-\alpha/2}$ and $-u_{1-\alpha/2}$ correspond to a unilateral test for positive or negative growth, respectively, with significance level $\alpha/2$. For the significance levels required, there is a choice of critical values from the appropriate table of percentiles for the typical normal distribution, which in this case is 1.64. From Table 4.2, it can be seen that the traction systems of the 5000-4th trains had a trend of high failure growth, according to the general behavior of the electromechanical systems.

Sections 7.2.2 and 7.3.1.1 of IEC 61710 (2013) on multiple items use the following formula for iterative estimation of $\hat{\beta}$:

$$\frac{N}{\hat{\beta}} + \sum_{i=1}^{N} \ln t_i - \frac{N \sum_{j=1}^{k} T_j^\beta \ln T_j}{\sum_{j=1}^{k} T_j^\beta} = 0 \tag{4.2}$$

where N is the total number of failures accumulated in the test, k is the total number of items, t_i is the time (or km) to the ith failure ($i = 1, 2, \ldots, N$), and T_j is the total time (or km) of observation for item $j = 1, 2, \ldots, k$. Then, $\hat{\lambda}$ can be calculated as follows:

$$\hat{\lambda} = \frac{N}{\sum_{j=1}^{k} T_j^{\hat{\beta}}}. \tag{4.3}$$

TABLE 4.2

Results of the Trend Test U for the Set of Items in the System

System	U	$U_{critical}$	Trend
Traction 5000-4th	16.88	1.64	Very growing

TABLE 4.3

Results of System Parameter Estimation with the PLP Model

System	β	λ	Trend
Traction 5000-4th	1.235	1.87 E^{-06}	Very growing

TABLE 4.4

Results of Goodness-of-fit Test of the System with the PLP Model

System	C^2	$C^2_{0.90}(M)$	PLP Model
Traction 5000-4th	2.108	0.173	Rejected

The model obtained is for the expected accumulated number of failures up to time t:

$$E[N(t)] = \lambda t^\beta. \tag{4.4}$$

Iterative calculation of $\hat{\beta}$, followed by calculation of $\hat{\lambda}$, gives the results shown in Table 4.3.

The goodness-of-fit test given in IEC 61710 (2013) is the Cramér–von Mises statistic C^2, with $M = N$ and $T = T^*$ for testing completed based on time, and $M = N - 1$ and $T = T_N$ for tests completed to failure:

$$C^2 = \frac{1}{12M} + \sum_{j=1}^{M}\left[\left(\frac{t_j}{T}\right)^\beta - \left(\frac{2j-1}{2M}\right)\right]^2. \tag{4.5}$$

A critical value of $C^2_{0.90}(M)$ is selected, with a level of significance of 10% of the tabulated value. If C^2 exceeds the critical value $C^2_{0.90}(M)$, $C^2 > C^2_{0.90}(M)$, then the hypothesis that the PLP model fits the test data must be rejected. When applying the PLP model for multiple items, as described in its Sections 7.2.2 and 7.3.1.1 of the IEC 61710 (2013), the model hypothesis is rejected. As shown in Table 4.4, the PLP model was rejected in the test system, owing to the dispersion of failure data of the traction systems of each train.

Then the models are applied to each of the 36 items, in accordance with the provisions of IEC standards. The following formula from Section 7.2 of IEC 60605-6 (2007) is used (Laplace test):

$$U = \frac{\sum_{i=1}^{r} T_i - r\frac{T^*}{2}}{T^*\sqrt{\frac{r}{12}}} \tag{4.6}$$

where r is the total number of failures to be considered, T^* is the total time (or km) of testing, and T_i is the time (or km) accumulated at the ith failure. The results obtained are summarized in Table 4.5. For each traction system of the 5000-4th trains, the trend was an increasing number of failures, although some items did not present a trend.

TABLE 4.5

Summary of test results of trend U for each item in the system

Train	U	$U_{critical}$	Trend
1	3.28	1.64	Growing
2	0.17	1.64	Without trend
3	4.09	1.64	Growing
4	0.77	1.64	Without trend
5	4.02	1.64	Growing
6	3.53	1.64	Growing
7	2.48	1.64	Growing
8	3.27	1.64	Growing
9	−0.48	1.64	Without trend
10	8.19	1.64	Growing
11	4.73	1.64	Growing
12	4.02	1.64	Growing
13	5.75	1.64	Growing
14	1.84	1.64	Growing
15	6.27	1.64	Growing
16	1.50	1.64	Without trend
17	2.20	1.64	Growing
18	5.63	1.64	Growing
19	1.54	1.64	Without trend
20	−1.26	1.64	Without trend
21	−0.57	1.64	Without trend
22	2.75	1.64	Growing
23	2.06	1.64	Growing
24	2.19	1.64	Growing
25	4.68	1.64	Growing
26	3.60	1.64	Growing
27	1.79	1.64	Growing
28	3.20	1.64	Growing
29	1.60	1.64	Without trend
30	0.79	1.64	Without trend
31	2.88	1.64	Growing
32	1.72	1.64	Growing
33	3.21	1.64	Growing
34	2.81	1.64	Growing
35	0.93	1.64	Without trend
36	2.90	1.64	Growing

TABLE 4.6

Estimation of λ Constant of Items of Each Traction System of 5000-4th Trains

Train	λ
2	6.08 E^{-05}
4	4.71 E^{-05}
9	4.58 E^{-05}
16	5.83 E^{-05}
19	4.38 E^{-05}
20	4.22 E^{-05}
21	4.54 E^{-05}
29	4.57 E^{-05}
30	6.45 E^{-05}
35	5.40 E^{-05}

For each of the ten items without a trend of failure, the λ constant was estimated using the HPP model proposed in the standard IEC 60605-4 (2001), Section 5.1 (see Table 4.6). For tests completed by time (or distance) and repairable items:

$$\hat{Z} = \hat{\lambda} = \frac{r}{T^*} \qquad (4.7)$$

where r is the total number of failures to be considered in the test, and T^* is the total time (or km) of testing completed at failure. It should be noted that the high degree of dispersion of the constant z(km) obtained between the system items represents unexpected results in traction systems, trains, and operational contexts that are in theory equal.

Finally, the PLP model was applied to each of the 26 items with a trend of failures, according to IEC 61710 (2013), Sections 7.2.1 and 7.3.1.1:

$$S_1 = \sum_{j=1}^{N} \ln\left(\frac{T^*}{t_j}\right) \qquad (4.8)$$

where T^* is the total time (or km) of testing and t_j is the accumulated time (or km) to the jth failure. Then, unbiased estimates were calculated for $\hat{\beta}$ and $\hat{\lambda}$ for the completed test as follows:

$$\hat{\beta} = \frac{N-1}{S_1} \qquad (4.9)$$

$$\hat{\lambda} = \frac{N}{k(T^*)^\beta} \qquad (4.10)$$

where N is the total number of accumulated failures in the test, and k is the total number of items in the test. It is necessary to carry out a goodness-of-fit test in order to check whether the model of reliability for each item properly fits the operating data, according to Section 7.3.1.1 of IEC 61710 (2013). The statistic C^2 must be calculated with $M = N$ and $T = T^*$ for testing based on time (or km):

$$C^2 = \frac{1}{12M} + \sum_{j=1}^{M}\left[\left(\frac{t_j}{T}\right)^\beta - \left(\frac{2j-1}{2M}\right)\right]^2. \qquad (4.11)$$

A critical value of $C^2_{0.90}(M)$ is selected, with a level of significance of 10% of the tabulated value. If C^2 exceeds the critical value $C^2_{0.90}(M)$, $C^2 > C^2_{0.90}(M)$, then the hypothesis that the PLP model fits the test data must be rejected.

The model obtained is for the expected accumulated number of failures up to time t:

$$E[N(t)] = \lambda t^\beta \qquad (4.12)$$

and for failure intensity:

$$z(t) = \frac{d}{dt}E[N(t)] = \lambda\beta t^{\beta-1}. \qquad (4.13)$$

A summary of the results is presented in Table 4.7.

- The 23 items with a trend of failures could not be modeled using PLP. The PLP model for items with a failure trend generally fails because the TBF does not have an exponential distribution. The following sections will analyze the potential causes of TBF not having an exponential distribution.
- For three items with a trend of failures, the model was accepted by a NHPP with the PLP model.
- The standard does not collect an NHPP model alternative to PLP for the 23 items with a rejected model. The high rates of models rejected in previous studies have led to the development of multiple models that respond adequately to the results in operation, such as complex SP models and others; see Ruggeri (2006) and Weckman et al. (2001) for examples applied to transport fleets.

4.3 The Complex Nature of Failures

Reliability differences in identical systems operating in equal operating contexts were found in the 36 traction systems of 5000-4th trains, which had reliabilities with different increasing or constant trends.

TABLE 4.7

Estimation of β and λ of the PLP Model of Each Traction System of 5000-4th Trains

Train	λ	β	$C^2_{0,90}$ (M)	C^2	PLP Model
1	4.83 E^{-07}	1.34	0.173	0.403	Rejected
3	1.03 E^{-06}	1.30	0.173	0.872	Rejected
5	1.12 E^{-07}	1.45	0.173	0.563	Rejected
6	4.60 E^{-06}	1.18	0.173	0.828	Rejected
7	1.10 E^{-07}	1.43	0.173	0.216	Rejected
8	1.09 E^{-06}	1.28	0.173	0.386	Rejected
10	4.57 E^{-10}	1.84	0.173	2.021	Rejected
11	5.92 E^{-08}	1.50	0.173	0.503	Rejected
12	1.37 E^{-07}	1.40	0.173	0.664	Rejected
13	5.93 E^{-09}	1.65	0.173	0.579	Rejected
14	2.49 E^{-06}	1.20	0.173	0.176	Rejected
15	5.05 E^{-09}	1.65	0.173	1.915	Rejected
17	2.89 E^{-07}	1.34	0.173	0.130	No rejected
18	6.57 E^{-11}	1.95	0.173	0.410	Rejected
22	2.56 E^{-07}	1.37	0.173	0.401	Rejected
23	2.83 E^{-06}	1.21	0.173	0.075	No rejected
24	4.57 E^{-06}	1.15	0.173	0.428	Rejected
25	3.52 E^{-05}	1.02	0.173	2.657	Rejected
26	2.18 E^{-06}	1.23	0.173	0.631	Rejected
27	1.13 E^{-04}	0.96	0.173	0.511	Rejected
28	3.44 E^{-07}	1.33	0.173	0.444	Rejected
31	2.27 E^{-05}	1.05	0.173	0.878	Rejected
32	7.41 E^{-06}	1.12	0.173	0.242	Rejected
33	2.64 E^{-07}	1.37	0.173	0.413	Rejected
34	1.66 E^{-07}	1.40	0.173	0.099	No rejected
36	7.45 E^{-06}	1.15	0.1730	0.4504	Rejected

Initially, the analysis focused on the search for patterns in the TBF of the systems, observing a generalized trend in which several consecutive failures accumulate during short temporal periods, preceded and followed by long periods without accumulated failures. This phenomenon, termed "recurrent failures," is well known to those responsible for maintaining repairable systems; see examples in Hatton (1999) and Karanikas (2013).

Subjectively, there is a perception that a series of complex repairable systems (e.g., automobiles) manufactured identically have different reliabilities in practical use. In some cases, studies have corroborated this subjective perception with data showing such disparate values in operation. Indeed, it is usually observed that each item in a set of complex repairable systems does not behave as predicted by a simple reliability model (HPP, NHPP, or RP), and that identical repairable items do not give the same reliability values, with notable differences in some cases.

For this reason, research has been diversified for the development of further models that can adequately represent the reliability of repairable systems. Such investigations were initiated by Lewis (1964), with a branching Poisson process (BPP), and Cox (1972), with a modulated renewal process (MRP). Models of "imperfect repair" include the BP model of Brown and Proschan (1983), the BBS model of Block et al. (1985), the trend renewal process (TRP) of Lindqvist et al. (2003), the generalized renewal processes (GRP) with the concept of "virtual age" introduced by Kijima (1989), and the proportional intensity (PI) models; for comparisons of these models, see Jiang et al. (2005) and Peña (2006). A novel contribution to the GRP model has recently been made by Kaminskiy and Krivtsov (2015). Another approach when the data are correlated is based on frailty models (Peña and Hollander, 2004). At present, more than 100 different models have been developed (Ascher and Feingold, 1984; Pham and Wang 1996; Rigdon and Basu 2000; Guo et al. 2000; Rausand and Hoyland, 2004; Peña, 2006).

Models grouped under the concept of "imperfect," including those based on SP, try to model the data of repairable systems while taking into account that the repair has not necessarily been "same as new" (perfect repair), nor has it necessarily been "same as old" (minimal repair). Therefore, they include an additional parameter or parameters to modulate the state of restoration to which each repair leads to the repairable item.

In most of these models, the TBF is a random variable with dependent increments, that is, it manifests some degree of relation with another random variable, which may be the previous repair, a preventive maintenance intervention, environmental conditions, etc. If dependence on TBF is demonstrated with some random variable, it statistically invalidates the convenience of using SP models; HPP, HNPP, and RP.

The recurrent failures that a repairable system accumulates during its long operating life are due to different causes, the diagnosis and solution of which may prove difficult in complex systems. Here, TBF data for each system were analyzed, and the most notable results are presented in Table 4.8. A high concentration of TBF can be observed near the origin, which represents recurrent failures. After a certain distance near the repair has been traveled, the distribution of the failures decreases in density. In no case does it correspond to an exponential distribution. The distributions that best represent the TBF sets per system are the generalized logistics for the 5000-4th traction system.

The practical interpretation of these values is that there is a high probability that a failure will occur in the first phases of operation after a repair, after which the probability of failure decreases significantly. Any correlation in the TBF should be ruled out to ensure that the data are independent, and SP models should be applied. This aspect will be analyzed in detail in the following sections, with the application of multivariate analysis.

TABLE 4.8

Basic Parameters of the TBF of Electric Traction Systems of 5000-4th Trains

Parameter	Traction 5000-4th
Values	3,112
Mean (km)	18,343
Standard deviation (km)	27,029
Coefficient of variation	147%
Maximum density (km)	13,000
% Values < Maximum density	61.92%
Best distribution (K-S test)	Generalized logistic

Also, as the results showed an evident dispersion in the reliability of each item of a system, the analysis of the reliability of the repairable systems may be completed through multivariate technical applications, in order to facilitate decision-making by those in charge of the maintenance of a fleet and/or set of systems consisting of items with disparate reliabilities.

Therefore, recurrent failures are the origin of TBF not having an exponential distribution, depending on the number of repetitive situations of episodes of recurrent failures that each item accumulates during long periods of operation. Recurrent failures substantiate the differences in trend and reliability values between each system item, even if the items are constructively identical and operate in homogeneous contexts.

4.4 Three-Dimensional Graphical Representation of $z(t)$ of the Repairable System

This type of graphics allows the first qualitative identification of the differences between the $z(t)$ of the items under study of each repairable system. Tang and Xie (2002) proposed a dispersion diagram of the λ_i that are obtained, in order to observe and graphically analyze these differences.

An example of such a graph is shown in Figure 4.2, for the electric traction system of 36 trains of the 5000-4th series, showing ten items with no trend of failure and 26 items with an increasing trend of failures.

Figure 4.2 shows the differences in z (km) obtained for each traction system of 5000-4th-series trains. In the foreground are the z (km) variables corresponding to the traction systems of the 26 trains with increasing model PLP, while in the background are those corresponding to the ten trains with an HPP model, sorted in ascending order (from lowest λ_i to major λ_i). This graph also reveals the dispersion of the z (km) of the items and how the values diverge with distance. Up to about 700,000 km, all trains have z (km) within

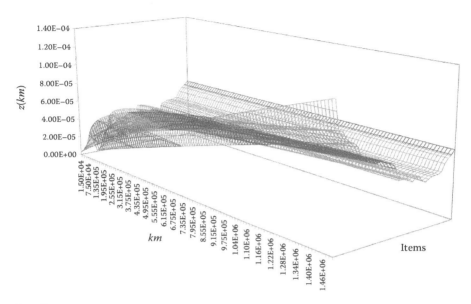

FIGURE 4.2
Three-dimensional graphical representation of the $z(km)$ of traction systems of 5000-4th trains.

a small range; however, at 1,500,000 km, trains with trends of increasing failures show $z(km)$ values that triplicate to trains with a trend constant of failures.

4.5 Potential Multivariate Applications for the Reliability Analysis of Repairable Systems

Multivariate analysis is a set of statistical methods whose purpose is to simultaneously analyze multivariate data sets, in the sense that there are several variables measured for each individual or object studied. These statistical methods help the analyst or researcher to make optimal decisions in the context in which they find themselves, taking into account the information available from the data set analyzed. They can be classified into three main groups.

- Dependence methods. These suppose that the analyzed variables are divided into two groups: the dependent variables and the independent variables. The purpose of dependence methods is to determine whether and in what way the set of independent variables affects the set of dependent variables.

- Methods of interdependence. These methods do not distinguish between dependent and independent variables; their purpose is to identify which variables are related, how they are related, and why.
- Structural methods. These assume that the variables are divided into two groups: the dependent variables and the independent variables. The purpose of these methods is to analyze not only how the independent variables affect the dependent variables, but also how the variables of the two groups are related to each other.

Of the existing statistical methods, eight were selected to be tested in the current reliability study of repairable systems, to potentially help and complement the previous modeling tests and results (HPP, NHPP, RP, etc.). Different statistical methods for multivariate analysis have been tested. These include:

- Correlations analysis
- Principal component analysis
- Factor analysis
- Cluster analysis
- Canonical correlation analysis
- Correspondence analysis
- Regression analysis
- Discriminant analysis

4.6 Correlations Analysis

Correlations analysis measures the strength of the linear relationship between two variables on a scale of −1 to +1. The greater the absolute value of the correlation, the stronger the linear relationship between the two variables. Its application to the reliability of the repairable systems can enable the discovery of dependencies or interdependencies in the TBF, as well as proper selection of the models of reliability that best adapt to the nature of the TBF.

The existence of correlation in the TBF should be ruled out to ensure that the data are independent. The potential variables that could be at the origin of the dependency of TBF in the traction systems were analyzed and evaluated, resulting in 12 variables being studied: train equipment, train alterations, mileage, years of operation, manufacturing date, manufacturing order, driving personnel, maintenance management system, maintenance plan, maintenance personnel, environmental conditions, and railway line. Applied Pearson correlation assays were used to rule out influence on TBF.

Of the 12 variables relating to traction systems of trains, only the results of the 2 variables with appreciable ranges are presented: date of manufacture and manufacturing order. The other ten variables had very similar and/or identical ranges, so that the correlation tests led directly to rejecting the dependency hypothesis with TBF. The influence of seasonality/temperature was ruled out, since the temperature in the tunnels was very stable throughout the year, with a maximum range between 20°C and 30°C. The results of correlations of TBF with context variables in the 5000-4th series trains are presented in Table 4.9.

TABLE 4.9

Correlations of the TBF with Context Variables in 5000-4th Trains

Train	TBF	TBF Mean	Manufacturing Order	Manufacturing Date
1	95	15,557	1	01/03/1993
2	97	16,040	2	01/03/1993
3	121	13,194	3	01/03/1993
4	77	20,950	4	01/03/1993
5	107	14,862	5	01/04/1993
6	103	16,331	6	01/04/1993
7	84	19,456	7	01/04/1993
8	99	16,468	8	01/04/1993
9	73	20,713	9	01/05/1993
10	126	12,973	10	01/05/1993
11	118	13,801	11	01/05/1993
12	73	22,275	12	01/05/1993
13	111	14,900	13	01/06/1993
14	67	24,126	14	01/06/1993
15	92	17,630	15	01/06/1993
16	96	17,091	16	01/06/1993
17	64	26,286	17	01/06/1993
18	79	20,124	18	01/06/1993
19	69	22,459	19	07/06/1993
20	67	22,336	20	01/07/1993
21	75	21,789	21	01/07/1993
22	84	19,061	22	01/07/1993
23	88	17,792	23	01/07/1993
24	60	25,078	24	01/07/1993
25	78	20,381	25	01/07/1993
26	90	16,571	26	01/07/1993
27	100	16,158	27	01/09/1993
28	62	25,918	28	01/09/1993
29	72	21,869	29	01/09/1993
30	100	15,416	30	01/10/1993
31	68	22,910	31	01/10/1993

(*Continued*)

TABLE 4.9 (*Continued*)

Correlations of the TBF with Context Variables in 5000-4th Trains

Train	TBF	TBF Mean	Manufacturing Order	Manufacturing Date
32	66	22,540	32	01/11/1993
33	79	19,259	33	01/11/1993
34	80	19,520	34	01/11/1993
35	85	18,518	35	01/12/1993
36	107	15,165	36	01/12/1993
Pearson correlation tests TBF			−0.3760	−0.2971
Pearson correlation tests TBF mean			0.3182	0.2416

In view of the Pearson correlation values, the influence on the TBF of each traction system of the studied variables could be ruled out. Bredrup et al. (1986) posited decades ago the existence of important differences between the $z(t)$ values obtained in identical systems, and analyzed the influence of nature of the failures: operational, hardware, software, etc.

A correlation study was also carried out on the TBF of each subsystem of the traction systems, and their influence on the TBF of each system, as well as when it was not possible to establish the origin of the failure in any specific part of the system, that is, when no repairs were carried out (only checks). No repairs were carried out were classified as subsystem "without apparent abnormality" (Table 4.10).

As shown in Table 4.11, there was no subsystem with significant correlation with total number of failures, except for "without apparent abnormality" (code 99) failures, and obviously the correlation must be very high, since this group accounts for 48.07% of the total number of failures.

TABLE 4.10

Legend of the Subsystems that Make up the Traction Systems of Trains

Code	Subsystem
11	Pantograph
12	Protections
13	Traction regulator
21	Relays and coils
22	Main switch
23	Power electronic
31	Contactors
32	50 Hz watch
33	Brake resistors
41	Electrical conduits
42	Cabin equipment
43	Traction motors
99	"Without apparent abnormality"

TABLE 4.11

Correlations of the Total Failures and Subsystems Failures in 5000-4th Trains

		TBF										
Train	Total	11	12	13	21	22	23	31	33	41	42	43	99
1	94	1	0	6	2	7	13	6	4	2	6	0	47
2	97	3	0	7	0	2	11	7	5	4	10	1	47
3	121	1	0	11	1	1	12	12	2	0	10	4	67
4	77	0	0	7	2	6	2	9	6	5	12	2	26
5	107	0	0	5	1	5	10	10	1	3	15	1	56
6	103	0	0	6	0	9	10	13	0	3	8	0	54
7	84	2	0	3	1	8	11	5	7	4	7	1	35
8	99	0	0	3	0	7	14	13	7	1	11	2	41
9	73	0	0	4	1	5	9	6	8	2	5	1	32
10	126	0	0	7	0	11	21	8	5	2	9	0	63
11	118	2	0	5	1	8	16	3	4	4	11	0	64
12	73	1	0	3	2	4	6	6	1	5	7	0	38
13	111	1	0	10	3	6	12	7	2	5	19	1	45
14	67	1	0	3	0	4	9	8	1	2	7	2	30
15	92	2	0	4	0	4	7	7	2	1	10	2	53
16	96	2	0	10	0	3	7	7	5	2	9	0	51
17	64	3	0	7	1	6	4	4	3	1	5	1	29
18	79	0	0	3	0	5	6	8	4	0	12	2	39
19	69	0	1	4	0	4	1	7	2	2	13	2	33
20	67	1	0	3	2	9	9	4	0	1	5	1	32
21	75	0	0	7	0	5	4	7	4	4	8	1	35
22	85	2	0	3	0	17	5	4	10	2	9	0	33
23	88	1	0	7	2	8	15	7	3	1	7	1	36
24	60	0	0	4	0	4	11	2	1	1	7	0	30
25	78	2	0	5	0	11	7	1	0	3	8	0	41
26	90	0	0	5	0	10	15	2	3	5	14	1	35
27	100	2	0	7	1	6	14	8	6	2	11	0	43
28	62	1	0	1	0	6	3	6	4	0	3	3	35
29	72	1	0	5	0	5	7	9	4	0	2	1	38
30	100	2	0	2	0	9	5	8	0	1	8	0	65
31	68	0	0	6	0	5	8	3	2	2	7	1	34
32	66	0	1	6	0	5	7	5	3	1	10	0	28
33	79	0	0	4	0	4	17	3	1	2	7	1	40
34	80	1	0	3	0	7	5	7	9	4	14	0	30
35	85	1	0	10	0	4	4	3	3	1	9	2	48
36	107	1	0	8	0	9	15	10	2	5	14	0	43
Pearson correlation tests total failures		0.13	−0.26	0.46	0.13	0.16	0.61	0.45	0.05	0.23	0.50	−0.10	0.84

TABLE 4.12

Basic Parameters of the TBF Groups of the 5000-4th Trains

Parameter	TBF Total	TBF without Abnormality	TBF with Repair
Values	3,112	1,496	1,616
Mean (km)	18,343	18,245	18,434
Standard deviation (km)	27,029	27,736	26,365
Coefficient of variation	147%	152%	143%
Maximum density (km)	13,000	13,000	13,000
% Values < Maximum density	61.92%	62.56%	61.38%
Best distribution (K-S test)	Generalized logistic	Three-parameter log-logistic	Three-parameter log-logistic

Therefore, it is also necessary to analyze whether the failures with repair "without apparent abnormality" have some degree of correlation with the recurrent failures, since, *a priori*, there seems to be a cause-effect relationship. That is, when a failure occurs and the repair is completed as "without apparent abnormality," the failure is likely to be repeated within a few kilometers. The method used for this analysis involves disaggregating and comparing the TBF in two groups: the first group corresponds to the TBF of the failures "without apparent abnormality," while the second is the remaining TBF in which a repair has been made (Table 4.12).

As shown in Figure 4.3, in the TBF of traction systems of the 5000-4th trains, a high concentration of TBF near the origin can be observed. This represents the recurrent failures, both in the TBF of the "without apparent

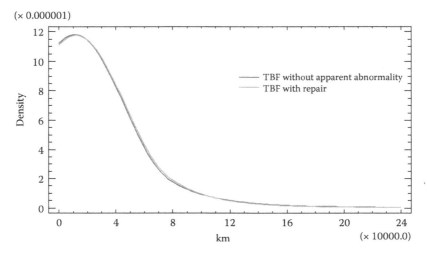

FIGURE 4.3
Density graph of the TBF groups of traction systems of 5000-4th trains.

abnormality" group and in those of the "repaired" group. The TBF densities of both groups are almost identical to each other, as well as to the total TBF of the system. In the traction systems of the 5000-4th series trains, it cannot be said that the recurrent failures have their origin in repairs "without apparent abnormality."

In view of the results obtained in the repairable system, it cannot be confirmed that there is a cause-effect relationship between a failure that, after review of the system, the maintenance technician classifies as "without apparent abnormality" and potential repetition of the failure, since the groups of TBF "without apparent abnormality" and "with repair" had almost identical distributions in all the systems.

Next, an attempt was made to set a critical TBF value for the identification of episodes with "accumulation of supposedly abnormal failures" and to typify these TBF as recurrent failures. It has been indicated that the distributions that best represent the TBF sets per system are the generalized logistics for the 5000-4th traction system (Table 4.8).

By adjusting the critical value of TBF for a lower tail area to <15%, the resulting TBF is very close to zero or negative. The interpretation of these results is that any small TBF, TBF → 0, is within the expected ranges of its distribution, so these TBF cannot be considered to be atypical.

Another approach is to try to set a critical number of failures per km interval for the identification of episodes with "accumulation of supposedly abnormal failures" and to typify these as recurrent failures. If, for example, in the traction systems of trains, the failures of each train are grouped in intervals of 15,000 km, it is possible to obtain a metric of "grouping" by interval such as that shown in the histogram of Figure 4.4.

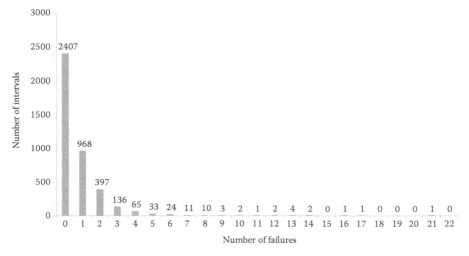

FIGURE 4.4
Histogram of accumulated failures in traction systems of 5000-4th trains.

The 15,000 km intervals have a distribution of the number of accumulated failures. In the case of the traction systems of the 5000-4th series trains, this is more suitably fitted to a three-parameter log-normal distribution. When the upper tail area limit was marked by 15%, the critical value was approximately two failures. In the analysis of the intervals with an accumulation of failures equal to or greater than two, corresponding to 693 out of a total of 4,068, it was not possible to observe any distribution depending on the following:

- The train
- The range in km
- Prior or subsequent intervals with accumulation (or not) of failures
- Number of failures accumulated in the interval

The graph in Figure 4.5 shows the failures of the traction systems of 32 5000-4th trains, with no apparent recognizable pattern.

In the correlation tests carried out, there was no statistically recognizable pattern in recurrent failures, there were episodes having an identical distribution with respect to km, and there was no apparent association between types of failures or repairs performed, as a consequence of which it is not possible to distinguish in practice between primary failures and supposed secondary failures. The concentration of failures in certain periods is qualitatively observable; it has not been possible to rule out the hypothesis that a

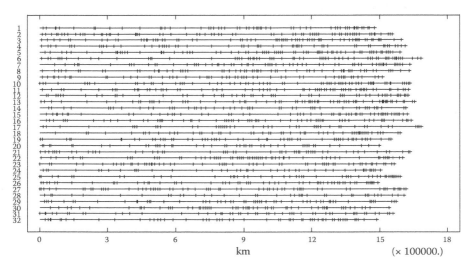

FIGURE 4.5
Graph of failures in 32 items of traction systems of 5000-4th trains.

failure f_i has a behavior independent of the near failure f_{i-1} that precedes it, and nor has it been possible to prove otherwise. Therefore, it cannot be ruled out that the HPP, NHPP, and RP models may be adequate to calculate and represent the reliability of the repairable system tested.

4.7 Principal Component Analysis

Principal component analysis is designed to extract k major components from a group of p quantitative variables. The main components are defined as a group of linear orthogonal combinations of X having the greatest variance. Determining the major components is often used to reduce the size of a group of predictive variables prior to their use in procedures such as multiple regression or cluster analysis. When variables are highly correlated, the first major components may be sufficient to describe most of the variability present.

In the study of the reliability of repairable systems, this method can be used to find a system or systems that are able to explain the behavior of the set of systems in the evolution of the expected cumulative number of failures $E[N(t)]$.

Principal component analysis of the system tested did not show any train that explains the evolution of the $E[N(t)]$ of the system as a whole. This result was expected, given the dispersion of the $E[N(t)]$ among the items in the system.

4.8 Factor Analysis

The factor analysis procedure is designed to extract m common factors from a group of p quantitative variables X. In many situations, a small number of common factors may be able to represent a large percentage of the variability in the original variables. The ability to express covariates between variables in terms of a small number of significant factors often leads to important questions about the data being analyzed.

Factor analysis may also be used in the study of the reliability of repairable systems to find a system or systems that are able to explain the behavior of the set of systems in the evolution of $E[N(t)]$.

The factor analysis of the system tested did not find any train that could explain the evolution of the $E[N(t)]$ of the system as a whole. Again, this result was expected, given the dispersion of the $E[N(t)]$ among the items in the system.

4.9 Cluster Analysis

The cluster analysis procedure is designed to group observations or variables into groups or clusters based on their similarities. The prime data for the procedure can be in either:

- n rows or cases, each containing the values of p quantitative variables
- n rows and n columns if observations are grouped, or p rows and p columns if variables are grouped, containing a measure of "distance" between all pairs of items

There are a number of different algorithms for generating groups. Some of the algorithms are agglomerative, starting with separate groups for each observation or variable and then joining them based on their similarity. Other methods begin with a set of "seeds," and tie other cases or variables to those seeds. In order to create clusters of observations or variables, it is important to have a measure of "closeness" or "similarity" such that similar objects can be joined. When observations are clustered, closeness is typically measured by the distance between observations in the p-dimensional space of variables. The test cluster analysis procedure contains three different metrics for measuring the distance between two objects, represented by x and y:

- Square Euclidean distance
- Euclidean distance
- "City block" distance

When variables are clustered, the distance is similarly defined, except that x and y represent the locations of two variables in the n-dimensional space of observations, and the sum is over observations rather than over variables. The methods tested are as follows:

- Agglomerative hierarchical methods: These methods start by putting each observation into a separate cluster. Clusters are linked, two at a time, until the number of clusters is reduced to a desired goal. At each stage, the clusters are paired according to their proximity: nearest neighbor, furthest neighbor, centroid, median, average linkage between groups (UPGMA), and Ward's method.
- k-means method: This method starts by identifying k objects as initial seeds for each cluster. Objects are attached to the nearest cluster.

Cluster analysis may be used in the study of the reliability of repairable systems for the generation of clusters (groups) of repairable systems that have a similar reliability. In many cases, the number of items to be analyzed

Potential Applications of Multivariate Analysis 131

is very high; the presentation of reliability results for more than 30 systems, for example, can be complex for maintenance management in business organizations. The creation of clusters in repairable systems by selecting outstanding variables of each item simplifies decision-making in such cases. Examples and procedures can be found in the following:

- Juarez et al. (2011); variables "dynamic parameters" applied to connection of multi-area power systems
- Yu and Chan (2012); variables "temperatures, power and flow" applied to operating performance of chiller systems
- Jaafar et al. (2012); variable "driving profiles" applied to the design of railway locomotives
- Shang and Wang (2015); variables "reliability, economy, and operational" applied to the power generation group
- Rastegari and Mobin (2016); variables "cost, frequency, and downtime" applied to a maintenance management system

In this study, "reliability" is proposed as a variable to simplify maintenance decisions in large fleets. For the creation of clusters in repairable systems without a trend, the values of λ_i calculated by the HPP reliability models for items without failure trends have been taken as variables. The number of clusters to be generated must be fixed before the test. In the presented results, three clusters have been created:

- High cluster: This cluster contains items with high λ that require different maintenance policies, increasing the consistency of preventive maintenance and requiring particular attention to the resolution of failures in corrective maintenance; these are the items with the worst reliability results.
- Middle cluster: This cluster contains items with intermediate λ that require regular maintenance policies, with consistent preventive maintenance and the resolution of failures in corrective maintenance.
- Low cluster: This cluster contains items with low λ that allow the policies of preventive and corrective maintenance to be relaxed, since these items have the best reliability results.

Cluster analysis was applied using the methods described above. Ward's method proved to be the most appropriate method for cluster creation, with a balanced number of items and results that were nearly independent of the distance metric used (square Euclidean, Euclidean, and city-block).

As an example, the results of applying this methodology to the 36 traction systems of 5000-4th trains are shown in Figure 4.6. The high cluster consists of seven items, the middle cluster 14, and the low cluster 15. This classification allows maintenance managers to more accurately adjust the different

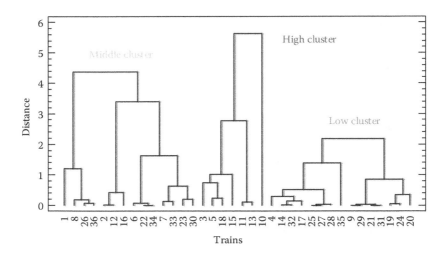

FIGURE 4.6
Three clusters of traction systems of 5000-4th trains generated by Ward's method.

maintenance strategies for groups of items with similar reliability behaviors, in order to use the available resources more efficiently.

Regardless of the method and metric used to create clusters of items with similar reliabilities in a repairable system, cluster analysis is considered a very useful tool to adjust decisions in maintenance strategies of large fleets, where it would be very complex, although desirable, to perform personalized maintenance for each item according to its reliability at each moment of operation.

4.10 Canonical Correlation Analysis

Canonical correlation analysis is designed to help identify associations between two sets of variables. This is done by finding linear combinations of the variables in the two sets that exhibit strong correlations. The pair of linear combinations with the strongest correlation form the first set of canonical variables. The second set of canonical variables is the pair of linear combinations that show the next strongest correlation, among all combinations that are not correlated with the first set. Frequently, a small number of pairs can be used to quantify the relationship that exists between the two sets.

This method should not be used for the creation of a single "type" individual calculated by the linear combination of the failures of n identical

Potential Applications of Multivariate Analysis 133

repairable systems, by not having with the necessary statistical support with respect to correct treatment of the TBF, and not contemplate the stochastic nature or dependence of the failures.

4.11 Correspondence Analysis

The correspondence analysis procedure creates a rows-and-columns map in a two-dimensional contingency table to superimpose related categories for row and column variables. However, no more than two or three dimensions are used to show the variability of inertia in the table. An important part of the output is the map of correspondences in which the distance between two categories is a measure of their similarity. No concrete application has been found for this method in the reliability analysis of repairable systems.

4.12 Regression Analysis

The regression analysis procedure is used to construct a statistical model that describes the impact of one or several quantitative factors $X_1 - X_i$, on a dependent variable Y. Multiple models and methods of regression analysis have been developed:

- Simple regression
- Multiple regression
- Logistic regression
- Negative binomial regression
- Nonlinear regression
- Polynomial regression
- Poisson regression
- Cox proportional hazards

Some of these methods have been used to model the reliability of repairable systems. The most commonly used models have been consolidated in different applied studies; see compendium in Liang (2011):

- Exponential smoothing model (ES)
- Moving average model (MA)

- Autoregressive integrated moving average process, Box-Jenkins ARIMA model
- Seasonal autoregressive integrated moving average process, Box-Jenkins SARIMA model

It is important to find the model that best fits the data, without needing to consider the reasons for the behavior of the TBF. An estimation of application of $E[N(t)]$ in different repairable systems by simple regression (least-squares method) is presented. Twenty-seven simple regression models were tested, with or without previous transformation of values of the X-axis (km) and of the Y-axis ($E[N(t)]$), with the coefficient $\beta_0 = 0$, since $E[N(0)] = 0$.

For each item, the regression model that best adapted to the data was selected, that is, the one with a higher of R^2 determination coefficient adjusted in a range from 0% to 100%. The goodness-of-fit test was integrated within the ANOVA (analysis of variance) model by decomposing the variability of the dependent variable Y into a sum of squares model of the error or residues. Of particular interest in this analysis is the test F and its associated P-value to test the statistical significance of the adjusted model. A small P-value (less than 0.05 at a significance level of 5%) indicates that a statistical relationship of the specified form exists between Y and X.

Table 4.13 shows the results of applying the simple regression models in $E[N(t)]$ to the traction systems of the 5000-4th trains. Models with limited adjustment to the failures were used. In nine items, the simple regression model that best adjusted was linear (without a trend in the failures), whereas in the tests for trends, the U statistics of these items were positive. Likewise, for each item, the goodness-of-fit test with a significance level of 5% admits between six and nine simple regression models, whose trend is not coincident in some cases, which questions the results and limitations of these simple models.

For the items with no trend in the failures, the most accepted and best fit simple regression model was the linear one, with $E[N(t)] = \beta_1 \text{km}$. As an example, the model for the traction system of train 20 of the 5000-4th series is shown in Figure 4.7. For items with an increasing trend in failures, the most accepted and best fit simple regression model was the square of x, with $E[N(t)] = \beta_1 \text{km}^2$. As an example, the model for the traction system of train 1 of the 5000-4th series is shown in Figure 4.8.

Many authors consider this type of model to be unorthodox, since they limit themselves to trying to adjust the data to a mathematical equation, without any explanation or formulation of the nature of the failures and/or their potential origin. This has been verified in the tests presented with these simple mathematical equations.

TABLE 4.13
Estimated Simple Regression Models of $E[N(t)]$ on 5000-4th Trains

Train	U Statistic Test Trend	Best Simple Regression	Model β Parameter	Trend of the Regression Model	Number of Models Accepted
1	Growing	Square-x	4.41 E^{-11}	Growing	6
2	Without trend	Linear	5.89 E^{-05}	Without trend	6
3	Growing	Square-x	5.18 E^{-11}	Growing	6
4	Without trend	Linear	4.60 E^{-05}	Without trend	6
5	Growing	Square-x	4.61 E^{-11}	Growing	6
6	Growing	Linear	4.92 E^{-05}	Without trend	6
7	Growing	Square-x	3.70 E^{-11}	Growing	6
8	Growing	Square-x	4.30 E^{-11}	Growing	6
9	Without trend	Linear	4.72 E^{-05}	Without trend	6
10	Growing	Square-x	3.83 E^{-11}	Growing	9
11	Growing	Square-x	4.68 E^{-11}	Growing	6
12	Growing	Linear	3.32 E^{-05}	Without trend	6
13	Growing	Square-x	3.99 E^{-11}	Growing	6
14	Growing	Linear	3.65 E^{-05}	Without trend	6
15	Growing	Square-x	2.91 E^{-11}	Growing	6
16	Without trend	Linear	5.52 E^{-05}	Without trend	6
17	Growing	Square-x	2.83 E^{-11}	Growing	6
18	Growing	Square-x	2.97 E^{-11}	Growing	6
19	Without trend	Logarithmic-y, square root-x	3.61 E^{-03}	Growing	9
20	Without trend	Linear	4.54 E^{-05}	Without trend	6
21	Without trend	Linear	4.66 E^{-05}	Without trend	6
22	Growing	Square of x	3.90 E^{-11}	Growing	9
23	Growing	Logarithmic-y, square root-x	3.84 E^{-03}	Growing	9
24	Growing	Linear	3.32 E^{-05}	Without trend	6
25	Growing	Linear	3.33 E^{-05}	Without trend	9
26	Growing	Logarithmic-y, square root-x	3.88 E^{-03}	Growing	9
27	Growing	Linear	5.58 E^{-05}	Without trend	9
28	Growing	Square of x	2.58 E^{-11}	Growing	6
29	Without trend	Linear	4.22 E^{-05}	Without trend	6
30	Without trend	Double square	4.73 E^{-09}	Without trend	6
31	Growing	Square of x	3.05 E^{-11}	Growing	6
32	Growing	Linear	3.93 E^{-05}	Without trend	6
33	Growing	Linear	4.15 E^{-05}	Without trend	6
34	Growing	Linear	4.46 E^{-05}	Without trend	6
35	Without trend	Double square	2.84 E^{-09}	Without trend	6
36	Growing	Logarithmic-y, square root-x	4.01 E^{-03}	Growing	9

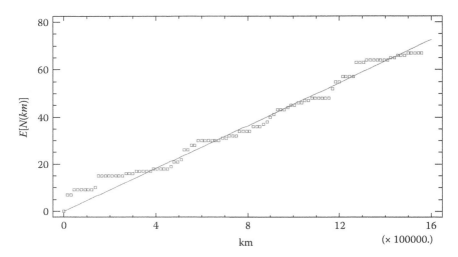

FIGURE 4.7
Linear simple regression model of train 20 of the 5000-4th series.

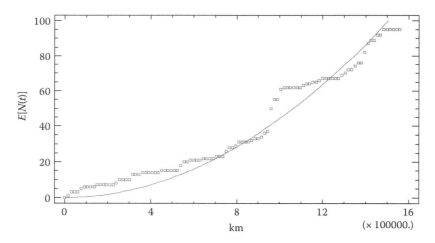

FIGURE 4.8
Square- x simple regression model of train 1 of the 5000-4th series.

4.13 Discriminant Analysis

The discriminant analysis procedure is designed to help distinguish between two or more data groups based on a group of p observed quantitative variables. It does so by constructing discriminant functions that are linear combinations of variables. The purpose of such an analysis is usually one or both of the following:

Potential Applications of Multivariate Analysis 137

- To be able to describe cases mathematically observed in a way that separates them into groups as best as possible
- To be able to classify new observations as belonging to one or other of the groups

No concrete application has been found to this method in the reliability analysis of repairable systems.

4.14 Conclusions

The TBF of items in repairable systems does not necessarily behave in a similar way. Even in what appears to be constructively identical repairable systems, the data may show a great dispersion in their trend and quantitative value of $E[N(t)]$.

The tests carried out showed that in all the repairable systems tested, periods of incremental failures preceded and followed periods without failure. This phenomenon is termed "recurrent failures," and is the reason why TBF does not have an exponential distribution and there are differences in reliability between constructively identical items and in homogeneous operational contexts.

In the periods of failure increments (recurrent failures), no statistically recognizable pattern was found. These episodes had identical distributions with respect to distance, with no apparent association between types of failures, resulting in primary and secondary failures being indistinguishable.

The items of a repairable system may show dispersed and in many cases divergent reliability values. Multivariate analysis methods may very well be useful to advance knowledge of the nature of the failures and the root causes of the differences in reliability between the items (Table 4.14).

TABLE 4.14

Multivariate Analysis Testing and Application to the Reliability of Repairable Systems

Method	Application
Correlations analysis	Dependency and/or interdependence in the TBF
Principal component analysis	Item or items that explain the reliability of a system
Factor analysis	Item or items that explain the reliability of a system
Cluster analysis	Generation of clusters of items that have a similar reliability
Canonical correlation analysis	Unidentified application
Correspondence analysis	Unidentified application
Regression analysis	Model the reliability of repairable systems (data oriented)
Discriminant analysis	Unidentified application

It is proposed that maintenance managers reorient their efforts to study the reliability of each item and not the set, and to implement mechanisms for the proactive detection of episodes of temporary failure increments in each item, applying differential maintenance in each case.

References

Anderson, G.B. and Peters, A.J. (1993), An overview of the maintenance and reliability of AC traction systems, *Proceedings of the Joint IEEE/ASME Railroad Conference*, pp. 7–15.

Ascher, H. and Feingold, H. (1984), *Repairable System Reliability*, Marcel Dekker, New York.

Attardi, L. and Pulcini, G. (2005), A new model for repairable systems with bounded failure intensity, *IEEE Transactions on Reliability*, Vol. 54, No. 4, pp. 572–582.

Bettini, G., Giansante, R. and Tucci, M. (2007), Forecasting fleet warranty returns using modified reliability growth analysis, *Proceedings of Annual IEEE Reliability and Maintainability Symposium RAMS'07*, pp. 350–355.

Block, H., Borges, W. and Savits, T. (1985), Age-dependent minimal repair, *Journal of Applied Probability*, Vol. 22, No. 2, pp. 370–385.

Bozzo, R., Fazio, V. and Savio, S. (2003), Power electronics reliability and stochastic performances of innovative AC traction drives: a comparative analysis, *2003 IEEE Bologna Power Tech Conference Proceedings*, Vol. 3, pp. 7–14.

Bredrup, E., Evensen, K., Helvik, B.E. and Swensen, A. (1986), The activity-dependent failure intensity of SPC systems—some empirical results, *IEEE Journal on Selected Areas in Communications*, Vol. 4, No. 7, pp. 1052–1059.

Brown, M. and Proschan, F. (1983), Imperfect repair, *Journal of Applied Probability*, Vol. 20, No. 4, pp. 851–859.

Chen, S.K., Ho, T.K. and Mao, B.H. (2007), Reliability evaluations of railway power supplies by fault-tree analysis, *IET Electric Power Applications*, Vol. 1, No. 2, pp. 161–172.

Cox, D.R. (1972), The statistical analysis of dependencies in point process, in Lewis, P.A.W. (Ed.), *Symposium on Point Processes*, Wiley, New York, pp. 55–66.

Crow, L.H. (1975), Reliability analysis for complex, repairable systems, Technical Report No. 138, AMSAA, Aberdeen, MD.

EN 50126-1 (1999), *Railway Applications—The Specification and Demonstration of Reliability. Availability, Maintainability, and Safety (RAMS)-Part, 1*. European Committee for Electrotechnical Standardization (CENELEC), Brussels.

Guo, R., Ascher, H. and Love, E. (2000), Generalized models of repairable systems—a survey via stochastic processes formalism, *ORiON*, Vol. 16, No. 2, pp. 87–128.

Hatton, L. (1999), Repetitive failure, feedback and the lost art of diagnosis, *Journal of Systems and Software*, Vol. 47, Nos. 2–3, pp. 183–188.

IEC 60300-3-1 ed2.0 (2003), *Dependability Management—Part 3-1: Application Guide—Analysis Techniques for Dependability—Guide on Methodology*, International Electrotechnical Commission (IEC), Geneva.

IEC 60300-3-5 ed1.0 (2001), *Dependability Management—Part 3-5: Application Guide—Reliability Test Conditions and Statistical Test Principles*, International Electrotechnical Commission (IEC), Geneva.

IEC 60605-4 ed2.0 (2001), *Equipment Reliability Testing—Part 4: Statistical Procedures for Exponential Distribution—Point Estimates, Confidence Intervals, Prediction Intervals and Tolerance Intervals*, International Electrotechnical Commission (IEC), Geneva.

IEC 60605-6 ed3.0 (2007), *Equipment Reliability Testing—Part 6: Tests for the Validity and Estimation of the Constant Failure Rate and Constant Failure Intensity*, International Electrotechnical Commission (IEC), Geneva.

IEC 61703 ed1.0 (2001), *Mathematical Expressions for Reliability, Availability, Maintainability and Maintenance Support Terms*, International Electrotechnical Commission (IEC), Geneva.

IEC 61710 ed2.0 (2013), *Power Law Model—Goodness-of-Fit Tests and Estimation Methods*, International Electrotechnical Commission (IEC), Geneva.

Jaafar, A., Sareni, B. and Roboam, X. (2012), Clustering analysis of railway driving missions with niching, *COMPEL—The International Journal for Computation and Mathematics in Electrical and Electronic Engineering*, Vol. 31, No. 3, pp. 920–931.

Jiang, S.T., Landers, T.L. and Rhoads, T.R. (2005), Semi-parametric proportional intensity models robustness for right-censored recurrent failure data, *Reliability Engineering and System Safety*, Vol. 90, No. 1, pp. 91–98.

Juarez, C., Messina, A.R., Castellanos, R. and Espinosa-Perez, G. (2011), Characterization of multimachine system behavior using a hierarchical trajectory cluster analysis, *IEEE Transactions on Power Systems*, Vol. 26, No. 3, pp. 972–981.

Kaminskiy, M. and Krivtsov, V. (2015), Geometric G1-renewal process as repairable system model, *Proceedings of Annual IEEE Reliability and Maintainability Symposium*, RAMS, Palm Harbor, FL, pp. 1–6.

Karanikas, N. (2013), Using reliability indicators to explore human factors issues in maintenance databases, *International Journal of Quality & Reliability Management*, Vol. 30, No. 2, pp. 116–128.

Kijima, M. (1989), Some results for repairable systems with general repair, *Journal of Applied Probability*, Vol. 26, No. 1, pp. 89–102.

Lewis, P. (1964), A branching Poisson process model for the analysis of computer failure patterns, *Journal of the Royal Statistical Society Series B (Methodological)*, Vol. 26, No. 3, pp. 398–456.

Liang, Y. (2011), Analyzing and forecasting the reliability for repairable systems using the time series decomposition method, *International Journal of Quality & Reliability Management*, Vol. 28, No. 3, pp. 317–327.

Lindqvist, B.H., Elvebakk, G. and Heggland, K. (2003), The trend-renewal process for statistical analysis of repairable systems, *Technometrics*, Vol. 45, No. 1, pp. 31–44.

Luo, M., Wu, M.L. and Wang, X.Y. (2010), Study on reliability test for brake control execution unit of rail transit vehicle, *2010 International Conference on E-Product, E-Service and E-Entertainment (ICEEE)*, pp. 1–4.

Panja, S.C. and Ray, P.K. (2007), Reliability analysis of track circuit of Indian railway signalling system, *International Journal of Reliability and Safety*, Vol. 1, No. 4, pp. 428–445.

Peña, E.A. (2006), Dynamic modeling and statistical analysis of event times, *Statistical Science*, Vol. 21, No. 4, pp. 487–500.

Peña, E.A. and Hollander, M. (2004), Models for recurrent events in reliability and survival analysis, in Soyer, T., Mazzuchi, T. and Singpurwalla, N. (Eds), *Mathematical Reliability: An Expository Perspective*, Kluwer Academic Publishers, Dordrecht, pp. 105–123.

Pham, H. and Wang, H. (1996), Imperfect maintenance, *European Journal of Operational Research*, Vol. 94, No. 3, pp. 425–438.

Rastegari, A. and Mobin, M. (2016), Maintenance decision making, supported by computerized maintenance management system, *IEEE 2016 Annual Reliability and Maintainability Symposium (RAMS)*, pp. 1–8.

Rausand, M. and Hoyland, A. (2004), System Reliability Theory: Models, Statistical Methods, *and Applications*, 2nd ed., Wiley, New York.

Rigdon, S.E. and Basu, A.P. (2000), *Statistical Methods for the Reliability of Repairable Systems*, Wiley, New York.

Ruggeri, F. (2006), On the reliability of repairable systems: methods and applications, *Proceedings of Progress in Industrial Mathematics at ECMI 2004*, pp. 535–553.

Sagareli, S. (2004), Traction power systems reliability concepts, *ASME/IEEE 2004 Joint Rail Conference*, pp. 35–39.

Shang, L. and Wang, S. (2015), Application of the principal component analysis and cluster analysis in comprehensive evaluation of thermal power units, *5th International Conference on Electric Utility Deregulation and Restructuring and Power Technologies (DRPT)*, Changsha, pp. 2769–2773.

Tang, L.C. and Xie, M. (2002), A simple graphical approach for comparing reliability trends of different units in a fleet, *Proceedings of Annual IEEE Reliability and Maintainability Symposium*, Seattle, WA, pp. 40–43.

Weckman, G.R., Shell, R.L. and Marvel, J.H. (2001), Modeling the reliability of repairable systems in the aviation industry, *Computers & Industrial Engineering*, Vol. 40, Nos. 1–2, pp. 51–63.

Yongqin, H. and Xishi, W. (1996), The reliability and performability of a repairable and degradable ATP system, *Vehicular Technology Conference, 1996, Mobile Technology for the Human Race, IEEE 46th*, Atlanta, GA, Vol. 3, pp. 1609–1612.

Yu, F.W. and Chan, K.T. (2012), Assessment of operating performance of chiller systems using cluster analysis, *International Journal of Thermal Sciences*, Vol. 53, pp. 148–155.

5

Phased Mission Systems—Modeling and Reliability

Kanchan Jain
Panjab University, Chandigarh

Isha Dewan
Indian Statistical Institute

Monika Rani
Defence Research Development Organisation

CONTENTS
5.1 Introduction 141
5.2 System Reliability Function When Subsystem Operational Times Are Unobservable 146
5.3 Study of System Operational Time for Exponential Life Times 148
    5.3.1 Subsystems with Different Number of Nonidentical Components 149
    5.3.2 Subsystems with Different Number of Identical Components 150
    5.3.3 Subsystems with Same Number of Identical Components 151
    5.3.4 Subsystems with Same Number of Nonidentical Components 152
5.4 Estimation of Parameter for Unobservable Subsystem Operational Times 153
    5.4.1 Subsystems with Different Number of Identical Components .. 154
    5.4.2 Subsystems with Equal Number of Identical Components..... 155
5.5 Simulations for Estimation of λ 157
5.6 Conclusion and Future Work 160
References... 160

5.1 Introduction

All complex systems can be considered as a collection of subsystems where each subsystem does its designated job. An automobile consists of several mechanical, electrical, and automated subsystems. Such systems are called phased mission (PM) systems. The system works when the subsystems work sequentially in several phases to complete the job for which it was designed. Another example of a PM system is an aircraft flight—different phases of the flight consist of takeoff, the ascent of the aircraft, altered flight due to interfaces, descent, and finally the landing. A boiling water reactor could face a "loss of coolant" accident. The corrective action in this case consists of three phases—cooling of initial core process, suppression core cooling, and removal of residual heat. Computer processing units, communication satellites, and satellite launchers are other examples of PM systems.

Defense equipment such as missiles and rockets is a PM system. The mission is considered successful only if each of the phases is successful. Each stage/phase of a multistage rocket involves an engine and a propellant. One of the many staging schemes for rockets is parallel staging scheme. In the first stage, the initial thrust exerted by the booster engines propels the entire rocket upward. The engine gets detached and falls off after the consumption of entire fuel. As a result, the rocket gets smaller and then the motor of the second stage fires. This continues till the last-stage motor burns to completion. The Remote Controlled Electronic Exploder ATLAS 400 RC used to fire electronic detonator is another example of such a system.

In reliability engineering, it is extremely important that the system performs its task satisfactorily. Many methods have been devised for finding the reliability of PM systems. The earliest of these was based on fault tree analysis and was given by Esary and Ziehms (1975). Phased algebra rules were derived by Dazhi and Xiaozhong (1989), Kohda et al. (1994), and Somani and Trivedi (1994). The techniques of phased algebra and fault tree were used by La Band and Andrews (2004) for determining the unreliability in individual phases. Other contributors in this area are Tillman et al. (1978), Alam and Al-Saggaf (1986), Alam et al. (2006), Altschul and Nagel (1987), Kim and Park (1994), Mura and Bondavalli (1999), Ma and Trivedi (1999), Zang et al. (1999), Meshkat (2000), Xing and Dugan (2002), Xing (2002, 2007a,b), Reay and Andrews (2002), Andrews and Beeson (2003), Tang and Dugan (2006), Trivedi (2006), Prescott et al. (2008, 2009), Mo (2009a,b), Remenyte-Prescott et al. (2010), Reed et al. (2011), He Hua-Feng et al. (2014), Levitin et al. (2012, 2013a,b, 2014a,b), and Mokhtarpour and Stracener (2015).

Two different phase-type systems have been illustrated in Figures 5.1 and 5.2. Figure 5.1 consists of k subsystems, where the ith subsystem consists of m_i subsystems in parallel. For the system to function effectively, each of the k subsystems must function. Since the components of the subsystem are in parallel, the first subsystem functions if at least one of the m_1 components is functioning, the second subsystem functions if at least one of m_2 components is functioning, and so on. Figure 5.2 consists of k subsystems where the ith subsystem consists of n_i subsystems in series. Since the components of the subsystem are in series, the first subsystem functions if each one of the n_1 components is functioning, the second subsystem functions

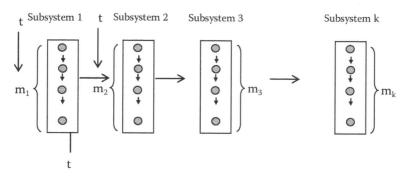

FIGURE 5.1
Depiction of the system with parallel configuration.

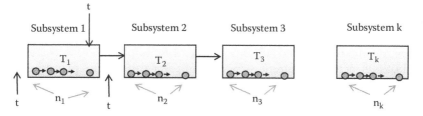

FIGURE 5.2
Depiction of the system with series configuration.

if each one of n_2 components is functioning, and so on. For such a system to complete its task, each one of $n_1 + n_2 + \cdots + n_k$ components must function successfully.

We wish to find the reliability of PM systems. The system operational time is always observable. However, the operational times of the k subsystems may be observable or unobservable. We give an example of both these cases. In defense applications, often electronic systems consisting of sequential phases/subsystems are encountered. An electronic safe/arm device has three phases: arming, charging, and finally firing. Their subsystems can be looked upon as phases of a mission with observable operational times, the time durations during which a system performs its intended task. However, there might be situations where the operational times of the subsystems may be unobservable. For example, missiles travel in different atmospheres before they inflict the desired damage on the target. Their trajectory is covered in different phases with some task assigned to each phase. The operational time taken by the missile through different phases may not be observable in some situations. In case of precision guided missiles, the trajectory is continuously changed for correct detection of target in its last phase. Hence, the operational time in the final phase will be unobservable.

The reliability of the two phase-type systems consisting of k (≥2) subsystems (illustrated in Figures 5.1 and 5.2) has been derived by assuming the following:

- Operational time of system is observable.
- Operational times of subsystems may be observable or unobservable.
- The lifetimes of all the components in the subsystems are independent.
- The subsystems operate independently.
- The components in a subsystem are in a parallel/series network.

The reliability estimation of phase-type systems when the subsystems consist of components in a parallel network with observable or unobservable operational times has been discussed in detail by Rani, Dewan, and Jain (2014) and Rani, Jain, and Dewan (2015). They gave results for distribution of the system operational time when the components within the subsystems are identically distributed as exponential but with different exponential distributions across different subsystems. The estimation procedure was described and estimates obtained through simulations. The system under study is assumed to consist of k (≥2) subsystems wherein lifetimes of components in different subsystems are independent and the components in each subsystem are in a parallel network. Let

- m_i be the number of components in the ith subsystem;
- X_{ij} denote the lifetime of the jth component in the ith subsystem for j = 1, 2, 3, ..., m_i;
- $T_i = \max(X_{i_1}, ..., X_{im_i})$ be the operational time of the ith subsystem for i = 1, 2, ..., k;
- $T = \sum_{j=1}^{k} T_j$ denote the operational time of the system;
- $F_z(\cdot)$ and $f_z(\cdot)$ be the cumulative distribution function (cdf) and probability density function (pdf) of a random variable Z.

In practice, m_is are 2 or 3. The number of standbys depends upon the cost and the criticality of the components. A highly critical component with less cost can have more standbys. If the subsystem operational times and system operational time are observable, then for general k, the pdf of system operational time T is

$$f_T(t) = \begin{cases} a_k^{-1} f_{T_1}(t), & 0 \leq t < t_1 \\ a_k^{-1} f_{T_1+T_1+\cdots+T_i}(t) & \sum_{j=1}^{i-1} t_j \leq t < \sum_{j=1}^{i} t_j \text{ for } i = 2,3,...,k \end{cases} \quad (5.1)$$

where

$$a_k = \int_0^{t_1} f_{T_1}(u)du + \sum_{i=2}^{k} \int_{\sum_{j=1}^{i-1} t_j}^{\sum_{j=1}^{i} t_j} f_{T_1+T_2+\cdots+T_i}(u)du. \qquad (5.2)$$

The constants a_k's are introduced so that $f_T(t)$ is a proper pdf.

The cdf of T is given by

$$F_T(t) = \begin{cases} 0, & t < 0 \\ a_k^{-1} A_1(t), & 0 \le t < t_1 \\ a_k^{-1} A_i(t), & \sum_{j=1}^{i-1} t_j \le t < \sum_{j=1}^{i} t_j \quad \text{for } i = 2,3,\ldots,k \end{cases} \qquad (5.3)$$

where

$$A_1(t) = F_{T_1}(t). \qquad (5.4)$$

For $i = 2, 3, \ldots, k$

$$A_i(t) = A_{i-1}(t_1 + t_2 + \cdots + t_{i-1}) + \int_{\sum_{j=1}^{i-1} t_j}^{t} f_{T_1+T_2+\cdots+T_i}(u)du. \qquad (5.5)$$

When the operational times of individual subsystems are not observable but the system operational time T can be observed, then for general k, the pdf of T is written as

$$f_T(t) = \begin{cases} 0, & t < 0 \\ b_k^{-1} f_{\sum_{j=1}^{k} T_j}(t), & 0 \le t < \sum_{j=1}^{k} t_j \end{cases} \qquad (5.6)$$

where

$$b_k = \int_0^{\sum_{j=1}^{k} t_j} f_{\sum_{j=1}^{k} T_j}(u)du. \qquad (5.7)$$

b_k's have been introduced so that $f_T(t)$ is a proper pdf.

The cdf of T is

$$F_T(t) = \int_0^t f_T(u)du = \begin{cases} 0, & t < 0 \\ b_k^{-1} F_{\sum_{j=1}^k T_j}(t), & 0 \leq t < \sum_{j=1}^k t_j \end{cases} \quad (5.8)$$

The reliability function of the system is

$$R_T(t) = 1 - F_T(t).$$

The expressions for the cdf of T have been found for k = 2 and 3 when the component lifetimes are independent and identical, each following exponential distribution with parameter λ. For an unequal number of identical components in two subsystems, estimates of λ and mean square errors (MSEs) for observable subsystem and system operational times and also for unobservable subsystem and observable system operational times have been reported through simulations.

Here we have considered the case when the subsystem components are in series but the operational life times of the subsystems are not observable. However, in some systems, the operational life times of the systems and all the subsystems may be observable. A study of such systems is under preparation and will be reported elsewhere.

In the present discussion, Section 5.2 presents the expressions for the cdf of the system lifetime and the reliability of the system for a series network with observable system operational time but unobservable subsystem operational times. In Section 5.3, expressions for the cdf and the reliability function have been derived when the lifetimes of components within subsystems follow exponential distribution with same or different parameters. The number of components in various subsystems have been taken to be equal or unequal. Section 5.4 discusses the problem of estimation for two cases namely subsystems with different number of identical components and subsystems with equal number of identical components. Simulations have been carried out for finding estimates of the unknown parameters of exponential distribution and corresponding MSEs for two possibilities in Section 5.5. Conclusions and future work have been reported in Section 5.6.

5.2 System Reliability Function When Subsystem Operational Times Are Unobservable

Let us consider a phase-type system with the components of the subsystems in series. Such a system has been depicted in Figure 5.2.

For j = 1, 2, 3, ..., n_i and i = 1, 2, 3, ..., k, let

X_{ij}: life time of the jth component in the ith subsystem;

$T_i = \min(X_{i1}, ..., X_{in_i})$: the unobservable operational time of the ith subsystem with t_i as a realization;

$T = \sum_{i=1}^{k} T_i$: the observable operational time of the system;

n_i: number of components in the ith subsystem.

In the sequel, the words "phase" and "subsystem" shall be used interchangeably. Same will apply to "lifetime" and "operational time."

For $t < t_1$, the control is in first subsystem and the assigned task is not yet complete;

$t_1 \leq t < t_1 + t_2$ means that subsystem 1 has completed its task and control is in subsystem 2.

In general, if $\sum_{j=1}^{i-1} t_j \leq t < \sum_{j=1}^{i} t_j$ for i = 3, ..., k, then subsystems 1, 2, ..., i – 1 have fulfilled their missions and control is in the ith subsystem.

When the operational times of individual subsystems are not observable whereas the system operational time is observable, we derive the expressions for the reliability function of system operational time T for a system consisting of k subsystems. In this case, the system operates till its mission is completed.

For k = 2, that is for two subsystems, the support of T gets truncated at $t_1 + t_2$ since the system operates till both the subsystems finish their assigned tasks and the cdf of T can be written as

$$F_T(t) = \begin{cases} 0, & t < 0 \\ c_2^{-1} F_{T_1+T_2}(t), & 0 \leq t < t_1 + t_2 \end{cases} \quad (5.9)$$

where

$$c_2 = P(T \leq t_1 + t_2) = \left(\int_0^{t_1+t_2} f_{T_1+T_2}(u) du \right). \quad (5.10)$$

The constant c_2 has been introduced so that the corresponding pdf $f_T(t)$ is a proper pdf.

It is to be noted that T_1 and T_2 are unobservable.

In general for k subsystems, the cdf of T takes the form

$$F_T(t) = \begin{cases} 0, & t < 0 \\ c_k^{-1} F_{k \atop \sum_{j=1} T_j}(t), & 0 \leq t < \sum_{j=1}^{k} t_j \end{cases} \quad (5.11)$$

where

$$c_k = \int_0^{\sum_{j=1}^{k} t_j} f_{k \atop \sum_{j=1} T_j}(u)\,du = F_{k \atop \sum_{j=1} T_j}\left(\sum_{j=1}^{k} t_j\right). \quad (5.12)$$

The reliability function of the system at time t is given by

$$R_T(t) = 1 - F_T(t) = \begin{cases} 1, & t < 0 \\ 1 - c_k^{-1} F_{k \atop \sum_{j=1} T_j}(t), & 0 \leq t < \sum_{j=1}^{k} t_j \end{cases} \quad (5.13)$$

In the next section, the expressions for the reliability function of the system are presented when lifetimes of components follow exponential distribution.

5.3 Study of System Operational Time for Exponential Life Times

This section includes the expressions for reliability function of the operational time of the system when the life times of the components follow exponential distribution and are independent of each other. It is assumed that the independent lifetimes X_{ij} of the jth component in the ith subsystem follow exponential distribution with parameter λ_i for $j = 1, 2, 3, ..., n_i$ and $i = 1, 2, 3, ..., k$. Hence

  i. the components within the system are iid but components in two different subsystems have different distributions.
  ii. since the components within the ith subsystem are in series, the life length of subsystem is $T_i = \min(X_{i1}, ..., X_{in_i})$ for $i = 1, 2, ..., k$, which follows exponential distribution with parameter $n_i \lambda_i$ for $i = 1, 2, ..., k$.

Phased Mission Systems—Modeling and Reliability

iii. subsystems are independent but not identically distributed.
iv. system lifetime is convolution of independent but not identical exponential random variables.

Hence for finding the pdfs of the sum of T_is, the distribution of convolutions of minimum of exponential random variables is used. In the following discussion, four possible cases arise.

5.3.1 Subsystems with Different Number of Nonidentical Components

We consider the case when $n_i \neq n_l$ and $\lambda_i \neq \lambda_l$ for $i \neq 1$ and $i, l = 1, 2, \ldots, k$, that is, subsystems have different distributions. The pdfs of $T = \sum_{j=1}^{i} T_j$ for $i = 2, 3, \ldots, k$ are given below and in all the expressions, it is assumed that $n_i\lambda_i \neq n_1\lambda_1$ for $i \neq 1$.
    For $k = 2$

$$f_{T_1+T_2}(t) = \int_0^t f_{T_2}(t-t_1)f_{T_1}(t_1)dt_1$$

$$= n_1 n_2 \lambda_1 \lambda_2 \left\{ \frac{e^{-t\lambda_2 n_2}}{\lambda_1 n_1 - \lambda_2 n_2} - \frac{e^{-t\lambda_1 n_1}}{\lambda_1 n_1 - \lambda_2 n_2} \right\}$$

$$= \sum_{i=1}^{2} n_i \lambda_i e^{-\lambda_i n_i t} \left[\prod_{j \neq i}^{2} \frac{n_j \lambda_j}{n_j \lambda_j - n_i \lambda_i} \right]. \tag{5.14}$$

For $k = 3$

$$f_{T_1+T_2+T_3}(t) = \int_0^t \int_0^{t-t_1} f_{T_3}(t-t_1-t_2)f_{T_2}(t_2)f_{T_1}(t_1)dt_2\, dt_1$$

$$= \sum_{i=1}^{3} n_i \lambda_i e^{-\lambda_i n_i t} \left[\prod_{j \neq i}^{3} \frac{n_j \lambda_j}{n_j \lambda_j - n_i \lambda_i} \right].$$

In general

$$f_{\sum_{j=1}^{k} T_j}(t) = \sum_{i=1}^{k} n_i \lambda_i e^{-\lambda_i n_i t} \left[\prod_{j \neq i}^{k} \frac{n_j \lambda_j}{n_j \lambda_j - n_i \lambda_i} \right].$$

The corresponding cdf of $\sum_{j=1}^{k} T_j$ is

$$F_{\sum_{j=1}^{k} T_j}(t) = \sum_{i=1}^{k}(1-e^{-\lambda_i n_i t})\left[\prod_{j\neq i}^{k} \frac{n_j \lambda_j}{n_j \lambda_j - n_i \lambda_i}\right].$$

Using (5.4),

$$c_k = \sum_{i=1}^{k}\left(1-e^{-\lambda_i n_i \sum_{l=1}^{k} t_l}\right)\left[\prod_{j\neq i}^{k} \frac{n_j \lambda_j}{n_j \lambda_j - n_i \lambda_i}\right].$$

Hence for $\lambda_i > 0$, using (5.11–5.13), the cdf of the system operational time and the system reliability function are as follows:

$$F_T(t) = \begin{cases} 0, & t < 0 \\ \dfrac{\sum_{i=1}^{k}(1-e^{-\lambda_i n_i t})\left[\prod_{j\neq i}^{k} \frac{n_j \lambda_j}{n_j \lambda_j - n_i \lambda_i}\right]}{\sum_{i=1}^{k}\left(1-e^{-\lambda_i n_i \sum_{l=1}^{k} t_l}\right)\left[\prod_{j\neq i}^{k} \frac{n_j \lambda_j}{n_j \lambda_j - n_i \lambda_i}\right]}, & 0 \leq t < \sum_{j=1}^{k} t_j; \end{cases} \quad (5.15)$$

$$R_T(t) = 1 - \dfrac{\sum_{i=1}^{k}(1-e^{-\lambda_i n_i t})\left[\prod_{j\neq i}^{k} \frac{n_j \lambda_j}{n_j \lambda_j - n_i \lambda_i}\right]}{\sum_{i=1}^{k}\left(1-e^{-\lambda_i n_i \sum_{l=1}^{k} t_l}\right)\left[\prod_{j\neq i}^{k} \frac{n_j \lambda_j}{n_j \lambda_j - n_i \lambda_i}\right]}, \quad 0 \leq t < \sum_{j=1}^{k} t_j. \quad (5.16)$$

5.3.2 Subsystems with Different Number of Identical Components

We assume that $n_i \neq n_l$ for $i \neq l$; $i, l = 1, 2, \ldots, k$ and $\lambda_i = \lambda$ for each i. Hence within a subsystem, the lifetimes of components within a subsystem are identically distributed. Since different subsystems have different number of components, subsystem life lengths T_is are nonidentical variables. Therefore, the subsystem lifetimes continue to be different.

The probability density function of $T_1 + T_2 + \cdots + T_k$ is as follows:

$$f_{\sum_{j=1}^{k} T_j}(t) = \lambda \sum_{i=1}^{k} n_i e^{-\lambda n_i t}\left(\prod_{j\neq i}^{k} \frac{n_j}{(n_j - n_i)}\right) \text{ for } \lambda > 0. \quad (5.17)$$

Phased Mission Systems—Modeling and Reliability

The corresponding cdf is as follows:

$$F_{\sum_{j=1}^{k} T_j}(t) = \sum_{i=1}^{k}\left(1-e^{-\lambda n_i t}\right)\left(\prod_{j \neq i} \frac{n_j}{(n_j - n_i)}\right) \quad (5.18)$$

and

$$c_k = \sum_{i=1}^{k}\left(1-e^{-\lambda n_i \sum_{j=1}^{k} t_j}\right)\left(\prod_{j \neq i} \frac{n_j}{(n_j - n_i)}\right). \quad (5.19)$$

Using (5.18) and (5.19) in (5.11), the cdf of the system operational time for $\lambda > 0$ is as follows:

$$F_T(t) = \begin{cases} 0, & t < 0 \\ \dfrac{\sum_{i=1}^{k}\left(1-e^{-\lambda_i n_i t}\right)\left[\prod_{j \neq i} \dfrac{n_j}{n_j - n_i}\right]}{\sum_{i=1}^{k}\left(1-e^{-\lambda n_i \sum_{j=1}^{k} t_j}\right)\left(\prod_{j \neq i} \dfrac{n_j}{n_j - n_i}\right)}, & 0 \leq t < \sum_{j=1}^{k} t_j \end{cases} \quad (5.20)$$

Equation (5.20) gives the system reliability function as

$$R_T(t) = 1 - \frac{\sum_{i=1}^{k}\left(1-e^{-\lambda n_i t}\right)\left[\prod_{j \neq i} \dfrac{n_j}{(n_j - n_i)}\right]}{\sum_{i=1}^{k}\left(1-e^{-\lambda n_i \sum_{j=1}^{k} t_j}\right)\left(\prod_{j \neq i} \dfrac{n_j}{(n_j - n_i)}\right)}, \quad 0 \leq t < \sum_{j=1}^{k} t_j. \quad (5.21)$$

5.3.3 Subsystems with Same Number of Identical Components

If $n_i = n_l$ for every i, l = 1, 2, …, k and $\lambda_i = \lambda$ for each i, then

a. across different subsystems component lifetimes are identically distributed.
b. subsystems are identical.
c. the system lifetime T is convolution of independent and identical distributed exponential random variables.

We can write

$$F_{\sum_{j=1}^{k} T_j}(t) = (1-e^{-\lambda n t}) - \sum_{i=1}^{k-1} \frac{(n\lambda t)^i}{i!} e^{-\lambda n t}. \quad (5.22)$$

Using (5.12),

$$c_k = \left(1 - e^{-\lambda n \sum_{j=1}^{k} t_j}\right) - \sum_{i=1}^{k-1} \frac{(n\lambda)^i}{i!} e^{-\lambda n \sum_{j=1}^{k} t_j} \left(\sum_{j=1}^{k} t_j\right)^i. \quad (5.23)$$

Hence using (5.22) and (5.23), the cdf of the system operational time T is given by

$$F_T(t) = \begin{cases} 0, & t < 0 \\ \dfrac{\left(1 - e^{-\lambda n t}\right) - \sum_{i=1}^{k-1} \dfrac{(n\lambda t)^i}{i!} e^{-\lambda n t}}{\left(1 - e^{-\lambda n \sum_{j=1}^{k} t_j}\right) - \sum_{i=1}^{k-1} \dfrac{(n\lambda)^i}{i!} e^{-\lambda n \sum_{j=1}^{k} t_j} \left(\sum_{j=1}^{k} t_j\right)^i}, & 0 \le t < \sum_{j=1}^{k} t_j. \end{cases}$$

(5.24)

This leads to the following expression for the reliability function of the system:

$$R_T(t) = 1 - \dfrac{\left(1 - e^{-\lambda n t}\right) - \sum_{i=1}^{k-1} \dfrac{(n\lambda t)^i}{i!} e^{-\lambda n t}}{\left(1 - e^{-\lambda n \sum_{j=1}^{k} t_j}\right) - \sum_{i=1}^{k-1} \dfrac{(n\lambda)^i}{i!} e^{-\lambda n \sum_{j=1}^{k} t_j} \left(\sum_{j=1}^{k} t_j\right)^i}, \quad 0 \le t < \sum_{j=1}^{k} t_j.$$

(5.25)

5.3.4 Subsystems with Same Number of Nonidentical Components

We study a system with k subsystems each containing the same number of nonidentical components, that is, $n_i = n_l$ but $\lambda_i \ne \lambda_l$ for i and l belonging to different subsystems. In this case, system life length is the sum of independent but not identical exponential random variables.

For $k = 2$,

$$f_{T_1 + T_2}(t) = n \sum_{i=1}^{2} e^{-n\lambda_i t} \prod_{j \ne i}^{2} \frac{\lambda_j}{(\lambda_j - \lambda_i)}. \quad (5.26)$$

On generalizing (5.26), we get

$$f_{\sum_{i=1}^{k} T_i}(t) = n \sum_{i=1}^{k} \lambda_i e^{-n\lambda_i t} \prod_{j \ne i}^{k} \frac{\lambda_j}{\lambda_j - \lambda_i}. \quad (5.27)$$

Using (5.11), (5.12) and (5.27), we have

$$F_{\sum_{i=1}^{k} T_j}(t) = n \sum_{i=1}^{k} \prod_{j \neq i}^{k} \frac{\lambda_j}{\lambda_j - \lambda_i}\left(1 - e^{-n\lambda_i t}\right) \qquad (5.28)$$

and

$$c_k = \sum_{i=1}^{k} \prod_{j \neq i}^{k} \frac{\lambda_j}{\lambda_j - \lambda_i}\left(1 - e^{-n\lambda_i \sum_{l=1}^{k} t_l}\right). \qquad (5.29)$$

Using (5.28) and (5.29), the cdf of the system operational time T can be written as

$$F_T(t) = \begin{cases} 0, & t < 0 \\[2mm] \dfrac{n \sum_{i=1}^{k} \prod_{j \neq i}^{k} \dfrac{\lambda_j}{\lambda_j - \lambda_i}\left(1 - e^{-n\lambda_i t}\right)}{\sum_{i=1}^{k} \prod_{j \neq i}^{k} \dfrac{\lambda_j}{\lambda_j - \lambda_i}\left(1 - e^{-n\lambda_i \sum_{l=1}^{k} t_l}\right)}, & 0 \leq t < \sum_{j=1}^{k} t_j. \end{cases} \qquad (5.30)$$

Hence the reliability function of the system is as follows:

$$R_T(t) = 1 - \frac{n \sum_{i=1}^{k} \prod_{j \neq i}^{k} \dfrac{\lambda_j}{\lambda_j - \lambda_i}\left(1 - e^{-n\lambda_i t}\right)}{\sum_{i=1}^{k} \prod_{j \neq i}^{k} \dfrac{\lambda_j}{\lambda_j - \lambda_i}\left(1 - e^{-n\lambda_i \sum_{l=1}^{k} t_l}\right)}, \quad 0 \leq t < \sum_{j=1}^{k} t_j. \qquad (5.31)$$

5.4 Estimation of Parameter for Unobservable Subsystem Operational Times

This section includes the maximum likelihood estimation of parameter of distribution of a series system operational time T. Subsystem life times T_1, T_2, ..., T_k are not observable here whereas system lifetime T is observable. We consider two cases. All components are iid exponential random variables with parameter λ but the number of units in each subsystem may be equal or unequal.

5.4.1 Subsystems with Different Number of Identical Components

We address the problem of estimating λ when it is assumed that $\lambda_i = \lambda$ for each i but $n_i \neq n_l$ for $i \neq l$, i, l = 1, 2, ..., k, that is, different subsystems have different number of components whose lifetimes are identically distributed as exponential with parameter λ.

Let $U_1, ..., U_m$ be a random sample of size m from $F_T(t)$ and $u_1, ..., u_m$ be the set of their observed values. For any $\lambda > 0$, the pdf of T is as follows:

$$f_T(t, \lambda) = c_k^{-1} \lambda \sum_{i=1}^{k} n_i N_i e^{-\lambda n_i t},$$

where $N_i = \prod_{\substack{j=1 \\ j \neq i}}^{k} \left(\frac{n_j}{n_j - n_i} \right)$ and c_k is given by (5.19).

The corresponding likelihood function can be written as

$$L = L(\lambda \mid u_1, ..., u_m) = \prod_{l=1}^{m} f_T(u_l, \lambda) = c_k^{-m} \lambda^m \left[\prod_{l=1}^{m} \left(\sum_{i=1}^{k} n_i N_i e^{-\lambda n_i u_l} \right) \right].$$

Hence

$$\frac{d \log L}{d \lambda} = \frac{m}{\lambda} + \sum_{l=1}^{m} \frac{\sum_{i=1}^{k} n_i N_i e^{-\lambda n_i u_l}}{\sum_{i=1}^{k} n_i N_i e^{-\lambda n_i u_l}} (-n_i u_l)$$

$$- m \left[\frac{\sum_{i=1}^{k} n_i N_i \left(\sum_{j=1}^{k} t_j \right) \left(e^{-\lambda n_i \sum_{j=1}^{k} t_j} \right)}{\sum_{i=1}^{k} N_i \left(1 - e^{-\lambda n_i \sum_{j=1}^{k} t_j} \right)} \right]$$

$$= \frac{m}{\lambda} - \sum_{l=1}^{m} \left[\frac{\sum_{i=1}^{k} n_i^2 N_i u_l e^{-\lambda n_i u_l}}{\sum_{i=1}^{k} n_i N_i e^{-\lambda n_i u_l}} \right]$$

$$- m \left[\frac{\sum_{i=1}^{k} n_i N_i \left[\sum_{j=1}^{k} t_j \right] \left(e^{-\lambda n_i \sum_{j=1}^{k} t_j} \right)}{\sum_{i=1}^{k} N_i \left(1 - e^{-\lambda n_i \sum_{j=1}^{k} t_j} \right)} \right]$$

$$J(\lambda) = \frac{-d^2 \mathrm{Log}\, L}{d\lambda^2}$$

$$= \frac{m}{\lambda^2} - \sum_{l=1}^{m} \left\{ \left(\frac{\sum_{i=1}^{k} n_i^2 N_i e^{-\lambda n_i u_l}}{\sum_{i=1}^{k} n_i N_i e^{-\lambda n_i u_l}} \right)^2 + \frac{\sum_{i=1}^{k} n_i^3 N_i e^{-\lambda n_i u_l}}{\sum_{i=1}^{k} n_i N_i e^{-\lambda n_i u_l}} \right\}$$

$$+ m \left\{ \frac{\sum_{i=1}^{k} N_i \left(1 - e^{-\lambda n_i \sum_{j=1}^{k} t_j}\right)\left(n_i \sum_{j=1}^{k} t_j\right)^2}{\sum_{i=1}^{k} N_i \left(1 - e^{-\lambda n_i \sum_{j=1}^{k} t_j}\right)} \right.$$

$$\left. - \left[\frac{\sum_{i=1}^{k} N_i \left(n_i \sum_{j=1}^{k} t_j\right)\left(1 - e^{-\lambda n_i \sum_{j=1}^{k} t_j}\right)}{\sum_{i=1}^{k} N_i \left(1 - e^{-\lambda n_i \sum_{j=1}^{k} t_j}\right)} \right]^2 \right\}.$$

In the following subsection, we consider finding the maximum likelihood estimator (MLE) of λ when the number of components in each subsystem is same and the lifetimes of each component are independent and identically distributed.

5.4.2 Subsystems with Equal Number of Identical Components

It is assumed that $n_i = n_l = n$ and $\lambda_i = \lambda_l = \lambda$ for $i, l = 1, 2, \ldots, k$. Hence the number of components in different subsystems is the same and the lifetimes of all the components across different subsystems are identically distributed. The pdf of system operational time T is as follows:

$$f_T(t) = \left(c_k^{-1}\right)(\lambda n)^k \left[e^{-\lambda n t} \frac{t^{k-1}}{(k-1)!} \right] \quad \text{for } \lambda > 0.$$

The corresponding likelihood function is as follows:

$$L = L(\lambda \mid u_1, \ldots, u_m) = \left(c_k^{-m}\right) \left[\frac{(\lambda n)^k}{(k-1)!} \right]^m \left[\prod_{l=1}^{m} u_l^{k-1} \right] e^{-\lambda n \sum_{l=1}^{m} u_l}$$

where

$$c_k = 1 - \sum_{i=0}^{k-1} \frac{\left(n\lambda \sum_{j=1}^{k} t_j\right)^i}{i!} e^{-\lambda n \sum_{j=1}^{k} t_j}.$$

This gives

$$\frac{d\log L}{d\lambda} = \frac{km}{\lambda} - n\sum_{l=1}^{m} u_l$$

$$+ \frac{\left\{\left[e^{-\lambda n \sum_{j=1}^{k} t_j}\right]\left[\sum_{i=0}^{k-1}\left(\frac{\lambda^{i-1}\left[n\left(\sum_{j=1}^{k} t_j\right)\right]^i}{(i-1)!} - \frac{\lambda^{i-1}\left[n\left(\sum_{j=1}^{k} t_j\right)\right]^{i+1}}{(i)!}\right)\right]\right\}}{1 - \sum_{i=0}^{k-1}\frac{\left[\lambda n\left(\sum_{j=1}^{k} t_j\right)\right]^i}{i!} e^{-\lambda n \sum_{j=1}^{k} t_j}}.$$

Solving $\frac{d\text{Log } L}{d\lambda} = 0$ shall provide us with the MLE of λ.

$$J(\lambda) = \frac{-d^2 \text{Log } L}{d\lambda^2}$$

$$= \frac{km}{\lambda^2} - \frac{(A-B)}{1 - \sum_{i=0}^{k-1}\frac{\left(n\lambda \sum_{j=1}^{k} t_j\right)^i}{i!} e^{-\lambda n \sum_{j=1}^{k} t_j}}$$

$$- \left\{\frac{\left[\left[e^{-\lambda n \sum_{j=1}^{k} t_j}\right]\left[\sum_{i=0}^{k-1}\left(\frac{\left[n\left(\sum_{j=1}^{k} t_j\right)\right]^i (\lambda)^{i-1}}{(i-1)!} - \frac{\left[n\left(\sum_{j=1}^{k} t_j\right)\right]^{i+1} (\lambda)^i}{(i)!}\right)\right]\right]^2}{1 - \sum_{i=0}^{k-1}\frac{\left(n\lambda \sum_{j=1}^{k} t_j\right)^i}{i!} e^{-\lambda n \sum_{j=1}^{k} t_j}}\right\}$$

where

$$A = e^{-\lambda n \sum_{j=1}^{k} t_j} \left[\sum_{i=0}^{k-1} \left(\frac{(\lambda)^{i-2}}{(i-2)!} \left(n \sum_{j=1}^{k} t_j \right)^i - \frac{(\lambda)^{i-1}}{(i-1)!} \left(n \sum_{j=1}^{k} t_j \right)^{i+1} \right) \right];$$

$$B = e^{-\lambda n \sum_{j=1}^{k} t_j} \left[\frac{n^{i+1}(\lambda)^{i-1}}{(i-1)!} \left(n \sum_{j=1}^{k} t_j \right)^{i+1} - \frac{(\lambda)^i}{i!} \left(n \sum_{j=1}^{k} t_j \right)^{i+2} \right].$$

In both the aforementioned cases, since $S(\lambda) = \frac{d\text{Log } L}{d\lambda} = 0$ cannot be solved analytically, the solution will be obtained by using the numerical method given by Newton-Raphson for finding MLE of λ and the values of estimates of λ are given in Section 5.5.

In the Newton Raphson method, assuming λ_0 to be the initial value of λ, the value of λ at the ith iteration can be found by using the relationship

$$\lambda_i = \lambda_{i-1} + \left[S(\lambda_{i-1}) \right] \left[J^{-1}(\lambda_{i-1}) \right].$$

This value is modified iteratively and the process continues until $|S(\lambda_j)| < \epsilon$ for some j and $\epsilon > 0$.

5.5 Simulations for Estimation of λ

The maximum likelihood estimation of parameter λ is discussed when the subsystem operational times are unobservable. The components' life times are assumed to be identically distributed as Exp (λ), $\lambda > 0$ and hence X_{ij}s are generated from Exp (λ) for $j = 1, 2, \ldots, n_i$ and $i = 1, 2, \ldots, k$. The number of components in different subsystems is assumed to be different.

T_i is computed as min $(X_{i_1}, X_{i_2}, \ldots, X_{in_i})$, $i = 1, 2, \ldots, k$.

In case of subsystems having different number of components with same distributions of lifetimes of components in different subsystems, the estimates of λ and corresponding MSEs are tabulated in Tables 5.1–5.3 for various choices of k, n_is, and λ. The considered sample sizes are m = 15, 20, and 25 since for expensive defense equipment, it is not feasible to replicate the systems a large number of times.

TABLE 5.1

Estimates of λ and MSEs for unequal number of identical components and m = 15

k	Sample size	n_i	λ	$\hat{\lambda}$	MSE
2	15	$n_1 = 3, n_2 = 5$	0.5	0.4665	0.0706
		$n_1 = 3, n_2 = 7$	0.7	0.6699	0.1314
		$n_1 = 3, n_2 = 9$	0.7	0.6688	0.1215
		$n_1 = 5, n_2 = 7$	0.7	0.6445	0.1392
3		$n_1 = 2, n_2 = 3, n_3 = 4$	0.8	0.7582	0.0955
		$n_1 = 2, n_2 = 5, n_3 = 6$	0.8	0.7613	0.0873
		$n_1 = 2, n_2 = 3, n_3 = 6$	0.8	0.7565	0.0925
		$n_1 = 3, n_2 = 5, n_3 = 6$	0.8	0.7618	0.0917
		$n_1 = 4, n_2 = 5, n_3 = 7$	0.8	0.7447	0.0963
4		$n_1 = 2, n_2 = 4, n_3 = 5, n_4 = 6$	0.8	0.7583	0.0619
		$n_1 = 2, n_2 = 4, n_3 = 5, n_4 = 7$	0.8	0.7621	0.0629
		$n_1 = 2, n_2 = 4, n_3 = 7, n_4 = 9$	0.8	0.7645	0.0585
		$n_1 = 3, n_2 = 5, n_3 = 7, n_4 = 9$	0.8	0.7641	0.0617

TABLE 5.2

Estimates of λ and MSEs for unequal number of identical components and m = 20

k	Sample size	n_i	λ	$\hat{\lambda}$	MSE
2	20	$n_1 = 3, n_2 = 5$	0.5	0.4762	0.0488
		$n_1 = 3, n_2 = 7$	0.7	0.6665	0.0958
		$n_1 = 3, n_2 = 9$	0.7	0.6798	0.0918
		$n_1 = 5, n_2 = 7$	0.7	0.6516	0.1019
3		$n_1 = 2, n_2 = 3, n_3 = 4$	0.8	0.7677	0.0719
		$n_1 = 2, n_2 = 5, n_3 = 6$	0.8	0.7718	0.0656
		$n_1 = 2, n_2 = 3, n_3 = 6$	0.8	0.7651	0.0692
		$n_1 = 3, n_2 = 5, n_3 = 6$	0.8	0.7679	0.0717
		$n_1 = 4, n_2 = 5, n_3 = 7$	0.8	0.7585	0.0696
4		$n_1 = 2, n_2 = 4, n_3 = 5, n_4 = 6$	0.8	0.7729	0.0458
		$n_1 = 2, n_2 = 4, n_3 = 5, n_4 = 7$	0.8	0.7679	0.0457
		$n_1 = 2, n_2 = 4, n_3 = 7, n_4 = 9$	0.8	0.7650	0.0467
		$n_1 = 3, n_2 = 5, n_3 = 7, n_4 = 9$	0.8	0.7701	0.0454

On the basis of Tables 5.1–5.3, the following can be concluded:

i. When the number of subsystems and number of components in the subsystems increase, the estimates of unknown parameter λ get closer to the assumed value of λ.

ii. With an increase in the number of subsystems and the number of components in the subsystems, MSE becomes smaller.

TABLE 5.3

Estimates of λ and MSEs for unequal number of identical components and m = 25

k	Sample Size	n_i	λ	$\hat{\lambda}$	MSE
2	25	$n_1 = 3, n_2 = 5$	0.5	0.4775	0.0397
		$n_1 = 3, n_2 = 7$	0.7	0.6731	0.0723
		$n_1 = 3, n_2 = 9$	0.7	0.6835	0.0718
		$n_1 = 5, n_2 = 7$	0.7	0.6651	0.0810
3		$n_1 = 2, n_2 = 3, n_3 = 4$	0.8	0.7654	0.0558
		$n_1 = 2, n_2 = 5, n_3 = 6$	0.8	0.7735	0.0519
		$n_1 = 2, n_2 = 3, n_3 = 6$	0.8	0.7739	0.0543
		$n_1 = 3, n_2 = 5, n_3 = 6$	0.8	0.7691	0.0559
		$n_1 = 4, n_2 = 5, n_3 = 7$	0.8	0.7683	0.0561
4		$n_1 = 2, n_2 = 4, n_3 = 5, n_4 = 6$	0.8	0.7728	0.0355
		$n_1 = 2, n_2 = 4, n_3 = 5, n_4 = 7$	0.8	0.7781	0.0356
		$n_1 = 2, n_2 = 4, n_3 = 7, n_4 = 9$	0.8	0.7817	0.0359
		$n_1 = 3, n_2 = 5, n_3 = 7, n_4 = 9$	0.8	0.7779	0.0365

TABLE 5.4

Estimates of λ and MSEs for Equal Number of Identical Components and k = 2

		m = 10		m = 15	
n	λ	$\hat{\lambda}$	MSE	$\hat{\lambda}$	MSE
2	0.9	0.9098	0.0080	0.8979	0.0051
2	0.6	0.6342	0.0024	0.5998	0.0025
2	0.5	0.4899	0.0022	0.5025	0.0016
3	0.7	0.7315	0.0039	0.6869	0.0041
3	0.5	0.4986	0.0044	0.5069	0.0026
4	0.6	0.5897	0.0017	0.6033	0.0043
4	0.5	0.4957	0.0027	0.4965	0.0013
4	0.4	0.3835	0.0020	0.4022	0.0023
4	0.3	0.3126	0.0012	0.2949	8.0834e-04
6	0.9	0.8742	0.0066	0.8922	0.0033
6	0.5	0.5120	0.0020	0.5017	0.0012
6	0.4	0.3844	0.0011	0.4159	0.0010
6	0.3	0.3073	0.0011	0.2939	5.9149e-04

It was also observed that for a large sample size, the estimates of unknown parameter λ were even closer to the assumed value of λ. This implies that the estimates are consistent. The MSEs were also smaller.

Table 5.4 displays the estimates of unknown parameter λ and corresponding MSE for various choices of λ when all subsystems have the same number of components whose lifetimes are identically distributed as exponential with parameter λ. In this table, n denotes the same number of components in each subsystem and k, the number of subsystems is two.

Similar conclusions as in Tables 5.1–5.3 can be drawn.

5.6 Conclusion and Future Work

In the current work and the work by Rani, Dewan, and Jain (2014) and Rani, Jain, and Dewan (2015), the component lifetimes are assumed to have exponential life distribution. Exponential random variables characterize no aging. Since this study was motivated from defense equipment which is in operation for very short duration, it was reasonable to assume that component lifetimes are exponential random variables. However, if components exhibit deterioration with time, then one can assume that life times have gamma or Weibull distributions. One would need to look at the reliability of phase-type systems under these distributions.

One can also look at systems with subsystems having components in more complex formations—different from series/parallel configuration.

References

Andrews J. D. and Beeson S. (2003). Birnbaum's measure of component importance for noncoherent systems, *IEEE Transactions on Reliability*, 52(2): pp. 213–219.

Alam M. and Al-Saggaf U. M. (1986). Quantitative reliability evaluation of repairable phased-mission systems using Markov approach, *IEEE Transactions on Reliability*, 35(5): pp. 498–503.

Alam M., Song M., Hester S. L. and Seliga T. A. (2006). Reliability analysis of phased mission systems: A practical approach, *Proceedings of the 52nd Annual Reliability & Maintainability Symposium*, Newport Beach, CA.

Altschul R. E. and Nagel P. M. (1987). The efficient simulation of phased fault trees, *Proceedings of IEEE Annual Reliability and Maintainability Symposium*, Philadelphia, PA: pp. 292–296.

Dazhi X. and Xiaozhong W. (1989). A practical approach for phased mission analysis, *Reliability Engineering and System Safety*, 25: pp. 333–347.

Esary J. D. and Ziehms H. (1975). Reliability analysis of phased missions, *Reliability and Fault Tree Analysis*, 13: pp. 213–236.

He H.F., Li J., Zhang Q. H. and Sun G. (2014). A data-driven reliability estimation approach for phased-mission systems, *Mathematical Problems in Engineering*, 2014, Article ID 283740, http://dx.doi.org/10.1155/2014/283740.

Levitin G., Xing L., and Amari S. V. (2012). Recursive algorithm for reliability evaluation of non-repairable phased mission systems with binary elements. *IEEE Transactions on Reliability*, 61(2): pp. 533–542.

Levitin G., Xing L., Amari S. V and Dai Y. (2013a). Reliability of nonrepairable phased-mission systems with common cause failures, *IEEE Transactions on Systems, Man and Cybernatics: Systems*, 43(4): pp. 967–978.

Levitin G., Xing L., Amari S. V. and Dai Y. (2013b). Reliability of non-repairable phased-mission systems with propagated failures, *Reliability Engineering and System Safety*, 119: pp. 218–228.

Levitin G., Xing L., Amari S. V. and Dai Y. (2014a). Cold vs. hot standby mission operation cost minimization for 1-out-of-N systems, *European Journal of Operational Research*, 234(1): pp. 155–162.

Levitin G., Xing L. and Yu S. (2014b). Optimal connecting elements allocation in linear consecutively-connected systems with phased mission and common cause failures, *Reliability Engineering & System Safety*, 130: pp. 85–94.

Kim K. and Park K. S. (1994). Phased mission system reliability under markov environment, *IEEE Transactions on Reliability*, 43: pp. 301–309.

Kohda T., Wada M. and Inoue K. (1994). A simple method for phased mission analysis, *Reliability Engineering and System Safety*, 45: pp. 299–309.

La Band R. and Andrews J. D. (2004). Phased mission modelling using fault tree analysis, *Proceedings of the Institution of Mechanical Engineers, Part E: Journal of Process Mechanical Engineering*, 218: pp. 83–91.

Ma Y. and Trivedi K. S. (1999). An algorithm for reliability analysis of phased-mission systems, *Reliability Engineering and System Safety*, 66(2): pp. 157–170.

Meshkat L. (2000). Dependency modeling and phase analysis for embedded computer based systems. Ph.D Dissertation, Systems Engineering, University of Virginia.

Mo Y. (2009a). New insights into the BDD based reliability analysis of phased mission systems, *IEEE Transactions on Reliability*, 58: pp. 667–678.

Mo Y. (2009b). Variable ordering to improve BDD analysis of phased mission systems with multimode failures, *IEEE Transactions on Reliability*, 58: pp. 53–57.

Mokhtarpour B. and Stracener J. T. (2015). Mission reliability analysis of phased-mission systems-of-systems with data sharing capability, *2015 Annual Reliability and Maintainability Symposium (RAMS)*, Palm Harbor, FL.

Mura I. and Bondavalli A. (1999). Hierarchical modelling and evaluation of phased mission systems, *IEEE Transactions on Reliability*, 48: pp. 360–368.

Prescott D. R., Remenyte-Prescott R., Reed S., Andrews J. D. and Downes C. G. (2008). A reliability analysis method using BDDs in phased mission planning, *Proceeding of IMech E, Journal of Risk and Reliability*, 223: pp. 27–39.

Prescott D. R., Andrews J. D. and Downes C. G. (2009). Multiplatform phased mission reliability modelling for mission planning, *Proceeding of Institute of Mechanical Engineers, Part O: Journal of Risk and Reliability*, 223: pp. 27–39.

Rani M., Jain K. and Dewan I. (2015). Estimation of reliability for parallel networked phased mission systems with unobserved subsystem operational times, *International Journal of Reliability, Quality and Safety Engineering*, 22(3), 15 pages.

Rani M., Dewan I. and Jain K. (2014). Estimation of parameters for phased mission systems-parallel network, *Advances in Reliability*, 1(1): pp. 1–6.

Reay K. A. and Andrews J. D. (2002). A fault tree analysis strategy using binary decision diagrams, *Reliability Engineering and System Safety*, 78(1): pp. 45–56.

Reed S., Andrews J. D., and Dunnett S. J. (2011). Improved efficiency in the analysis of phased mission systems with multiple failure mode components, *IEEE Transactions on Reliability*, 60(1): pp. 70–79.

Remenyte-Prescott R., Andrews J. D. and Chung P. W. H. (2010). An efficient phased mission reliability analysis for autonomous vehicles, *Reliability Engineering and System Safety*, 95: pp. 226–235.

Somani A. K. and Trivedi K. S. (1994). Boolean algebraic methods for phased mission system analysis, *Proceedings of ACM Sigmetrics, Conference on Measurement and Modeling of Computer Systems*: pp. 98–107.

Tang Z. and Dugan J. B. (2006). BDD based reliability analysis of phased mission systems with multimode failures, *IEEE Transactions on Reliability*, 55: pp. 350–360.

Trivedi K. S. (2006). *Probability and Statistics with Reliability, Queuing, and Computer Science Applications*, John Wiley & Sons, New York.

Tillman F. A., Lie C. H. and Hwang C. L. (1978). Simulation model of mission effectiveness for military systems, *IEEE Transactions on Reliability*, R-27: pp. 191–194.

Xing L. and Dugan J. B. (2002). Analysis of generalised phased mission system reliability, performance and sensitivity, *IEEE Transactions on Reliability*, 51(2): pp. 199–211.

Xing L. (2002). Dependability modeling and analysis of hierarchical computer-based systems. Ph.D. Dissertation, Electrical and Computer Engineering, University of Virginia.

Xing L. (2007a). Reliability importance analysis of generalized phased-mission systems, *International Journal of Performability Engineering*, 3(3): pp. 303–318.

Xing L. (2007b). Reliability evaluation of phased mission systems with imperfect fault coverage and common- cause failures, *IEEE Transactions on Reliability*, 56: pp. 58–68.

Zang X., Sun H. and Trivedi K. S. (1999). A BDD based algorithm for reliability analysis of phased mission systems, *IEEE Transactions on Reliability*, 48: pp. 50–60.

6

Bayesian Inference on General-Order Statistic Models

Aniket Jain
Acellere Software Pvt. Ltd

Biswabrata Pradhan
Indian Statistical Institute

Debasis Kundu
Indian Institute of Technology

CONTENTS

6.1	Introduction	163
6.2	Model Assumptions and Prior Selection	166
	6.2.1 Model Assumptions	166
	6.2.2 Prior Selection	167
6.3	Posterior Analysis of Different GOS Models	167
	6.3.1 Exponential GOS Model	167
	6.3.2 Weibull GOS Model	169
	6.3.3 Generalized Exponential GOS Model	170
6.4	Simulation Experiments	172
6.5	Data Analysis	176
6.6	Conclusion	180
Acknowledgment		180
Appendix		180
References		181

6.1 Introduction

Suppose there is a closed population of size N. In this chapter, we consider the estimation of N based on Type-I censored data. The problem is the following. Let $T_1, \ldots, T_N$ be a random sample of a positive random variable having the positive probability density function (PDF) at $x = f(x; \delta)$. In this case, N and δ both are unknown; moreover, δ may be vector valued also. Let T^* be a prefixed time, denoting the period of observations. Suppose we observe $0 < t_{(1)} < \cdots < t_{(r)} < T^*$ within the observation period T^*. There is no failure between $t_{(r)}$ and T^*. We would like to draw the inference on N and δ, based on the above Type-I censored sample.

This is known as the general-order statistics (GOS) model, and it has several applications in different areas. Consider the example of a population cited by Hoel (1968), where some members of a given population are exposed to disease or radiation at a given time. Let N be the number of individuals exposed to radiation. It is assumed that the times from exposure to detection of these individuals are random, and they are independent and identically distributed (i.i.d.) random variables, say $T_1, \ldots, T_N$, from a PDF $f(x;\delta)$. Based on the first r ordered sample until time point T^*, the problem is to estimate N and δ.

A similar problem can occur in software reliability, see, for example, Jelinski and Moranda (1972). Here one is interested in estimating the number of faults or bugs in a software from the initial failure times, $t_{(1)} < \cdots < t_{(r)}$, observed during an observation period T^*. Anscombe (1961) provided an interesting example of this model in estimating the sales of a company's product in a particular market. The main aim is to predict the average sales of the product in the future based on the information obtained during a short period after its penetration into the market. Osborne and Severini (2000) also considered the same problem to estimate the size N of a closed population based on the available observations up to a fixed length $T^* > 0$. They have considered the estimation of N under an exponential general-order statistic model.

Johnson (1962) and Hoel (1968) gave the methods for discriminating between two values of N based on the likelihood ratio and the sequential probability ratio test, respectively, when underlying lifetime distributions are completely known. Blumenthal and Marcus (1975) provided the maximum likelihood estimator (MLE) of N assuming the underlying probability distribution as exponential. Jelinski and Moranda (1972), Forman and Singpurwalla (1977), Meinhold and Singpurwalla (1983), Jewell (1985), Joe and Reid (1985), and Joe (1989) also considered this problem in the context of software reliability where the problem is to estimate the number of faults N in a software. The models proposed by them are the extensions of the model originally proposed by Jelinski and Moranda (1972), where the underlying lifetime distribution is exponential. Raftery (1987) considered estimation of the unknown population size N under general-order statistic model and adopted an empirical Bayes approach. In particular, Raftery (1987) considered single parameter Weibull and Pareto order statistic models, which mainly belong to the exponential family. Although the performance of point estimators was not satisfactory, the interval estimators can be obtained and they might be useful for practical purposes. The problem has a close resemblance to the problem of estimating n for a binomial random variable. Extensive work has been done in estimating n of a binomial population; see, for example, DasGupta and Herman (2005) in this respect.

Kuo and Yang (1995, 1996) also considered similar models. They have provided an interesting connection between GOS models and nonhomogeneous Poisson processes. They have also adopted Bayesian inference, but their problems of interests are slightly different than ours. They have mainly considered in detail the Bayesian prediction and model selection, not the estimation of N.

It may be mentioned that the estimation of N is a nontrivial problem. The point estimator of N (say $\widehat{N}_{MLE}$) obtained by maximizing the likelihood function has several unusual features. It is observed that $P\left(\widehat{N}_{MLE} = \infty\right) > 0$. Also it is well known that both the mean and median are biased estimators but in opposite directions. With a very high probability $\widehat{N}_{MLE}$ can take large values and it falls below the actual parameter value quite frequently. Furthermore, it is quite unstable, and a small change in the data can lead to a large change in $\widehat{N}_{MLE}$.

Our proposed method is purely Bayesian in nature mainly for two purposes. First of all, it avoids the problem of finding an estimator which is not finite. Second, although the exponential GOS model has been studied quite extensively by several authors, not much attention has been paid for other distributions. It is observed that if the lifetime distribution is not exponential, analytically it becomes a challenging problem in the frequentist setup. It seems that for many lifetime distributions, the implementation of the Bayesian analysis is quite straightforward. Here, we have considered several lifetime distributions, namely (i) exponential, (ii) Weibull, and (iii) generalized exponential models. Suitable theories and proper implementation procedures have been developed for point and highest posterior density (HPD) credible interval estimation of N and other unknown parameters.

The choice of prior plays an important role in any Bayesian inference problem. An independent Poisson prior has been assigned to N and for three different lifetime distributions, quite flexible priors to the unknown parameters of the distribution of T have been assumed. Based on the prior distributions and data, posterior distributions are obtained. All the estimates are obtained under the squared error loss (SEL) function. The Bayes estimators under the SEL function cannot be obtained explicitly. Hence, Markov Chain Monte Carlo (MCMC) technique has been used to compute the Bayes estimates and the associated credible intervals. Extensive simulation experiments have been performed to assess the effectiveness of the proposed methods. The performances are quite satisfactory. The analysis of one real data set has been presented to illustrate the proposed methods.

We have organized the remaining chapter as follows. The models and the priors have been presented in Section 6.2. In Section 6.3, the posterior analysis under different lifetime distributions has been provided. Monte Carlo simulation results have been presented in Section 6.4. In Section 6.5, we provide the analysis of a real data set. Finally, in Section 6.6, the conclusion has been provided.

6.2 Model Assumptions and Prior Selection

6.2.1 Model Assumptions

Suppose $T_1, \ldots, T_N$ is a random sample of a positive random variable with PDF $f(x;\delta)$, and cumulative distribution function (CDF) $F(x;\delta)$. Let the first r order statistics $t_{(1)} < \cdots < t_{(r)} < T^*$ be observed within the observation period T^*. The likelihood function is then given by

$$L(N,\delta \mid \text{data}) = \frac{N!}{(N-r)!}\left(\prod_{i=1}^{r} f(t_{(i)};\delta)\right)\left(1 - F(T^*;\delta)\right)^{N-r}, \quad N = r, r+1, \ldots \quad (6.1)$$

The problem is to estimate N and δ and we have assumed the following different parametric forms of $f(x;\delta)$.

Exponential model: It is the most commonly used lifetime distribution. Analytically, it is the most tractable lifetime distribution. In this chapter we have assumed the following PDF of an exponential distribution for $\lambda > 0$.

$$f_{EX}(t;\lambda) = \begin{cases} \lambda e^{-\lambda t} & \text{if} \quad t > 0 \\ 0 & \text{if} \quad t \leq 0. \end{cases} \quad (6.2)$$

Weibull model: Although the exponential distribution is used quite extensively as a lifetime distribution, it has a decreasing PDF and a constant hazard function. These are serious limitations for an exponential distribution. The Weibull distribution has two parameters: one shape parameter and one scale parameter. The presence of the shape parameter makes it a very flexible distribution. The Weibull distribution has a decreasing or an unimodal density function. If the shape parameter is less than or equal to one, it has a decreasing PDF. Otherwise, the PDF is an unimodal function. Furthermore, the hazard function also can take various shapes namely increasing, decreasing, or constant. It can be used quite successfully to analyze lifetime data. The Weibull distribution has the following PDF for $\alpha > 0$ and $\lambda > 0$;

$$f_{WE}(t;\alpha,\lambda) = \begin{cases} \alpha\lambda t^{\alpha-1} e^{-\lambda t^{\alpha}} & \text{if} \quad t > 0 \\ 0 & \text{if} \quad t \leq 0. \end{cases} \quad (6.3)$$

Generalized exponential model: Gupta and Kundu (1999) introduced the generalized exponential distribution which behaves very similarly as the Weibull or gamma distribution. For details, see the survey article by Nadarajah (2011). The generalized exponential distribution has the following PDF for $\alpha > 0$ and $\lambda > 0$:

$$f_{GE}(t;\alpha,\lambda) = \begin{cases} \alpha\lambda e^{-\lambda t}(1 - e^{-\lambda t})^{\alpha-1} & \text{if} \quad t > 0 \\ 0 & \text{if} \quad t \leq 0. \end{cases} \quad (6.4)$$

6.2.2 Prior Selection

A Poisson random variable X with mean μ (POI(μ)) has the following probability mass function (PMF);

$$P(X = i) = \frac{e^{-\mu}\mu^i}{i!}; \quad i = 0, 1, \ldots. \tag{6.5}$$

The PDF of a gamma random variable with the shape parameter $a > 0$ and the scale parameter $b > 0$ (GA(a,b)) is as follows:

$$f_{GA}(x;a,b) = \begin{cases} \dfrac{b^a}{\Gamma(a)} x^{a-1} e^{-bx} & \text{if } x > 0 \\ 0 & \text{if } x \le 0. \end{cases} \tag{6.6}$$

The following prior assumptions have been made. In all the three cases considered here: N follows (~) POI(θ). In case of exponential distribution $\delta = \lambda$, and it is assumed that $\lambda \sim$ GA(c,d). Moreover, N and λ are independently distributed. For two-parameter Weibull and two-parameter generalized exponential distributions, $\delta = (\alpha, \lambda)$, and in both the cases $\alpha \sim$ GA(a,b). The prior on λ is the same as before, and all priors are assumed to be independent.

6.3 Posterior Analysis of Different GOS Models

6.3.1 Exponential GOS Model

The likelihood function (6.1) for $N = r, r+1, \ldots$ and $\lambda > 0$, becomes

$$L_{EX}(N, \lambda \mid data) = \frac{N!}{(N-r)!} \lambda^r \left(e^{-\lambda\left(\sum_{i=1}^{r} t_{(i)} + (N-r)T^*\right)} \right). \tag{6.7}$$

Hence, based on the prior distributions of N and λ, the posterior distribution of N and λ, is as follows:

$$\pi_{EX}(N, \lambda \mid data) \propto \frac{\theta^N}{(N-r)!} \lambda^{c+r-1} \left(e^{-\lambda\left(\sum_{i=1}^{r} t_{(i)} + (N-r)T^* + d\right)} \right). \tag{6.8}$$

Let $M = N - r$, then the joint posterior distribution of M and λ is given by

$$\pi_{EX}(M, \lambda \mid data) \propto \frac{\theta^{M+r}}{M!} \lambda^{c+r-1} \left(e^{-\lambda\left(\sum_{i=1}^{r} t_{(i)} + MT^* + d\right)} \right). \tag{6.9}$$

If $g(M,\lambda)$ is a function of M and λ, then under SEL function

$$\hat{g}_B(M,\lambda) = E(g(M,\lambda)) = \sum_{m=0}^{\infty} \int_0^{\infty} g(m,\lambda)\pi_{EX}(m,\lambda\,|\,data)d\lambda \qquad (6.10)$$

is the Bayes estimate of $g(M,\lambda)$.

It is clear that (6.10) cannot be expressed in explicit form, hence we use the Monte Carlo simulation to approximate (6.10). First, we observe that the joint posterior density function (6.9) is given by

$$\pi_{EX}(M,\lambda\,|\,data) = \pi_{EX}(\lambda\,|\,M,data) \times \pi_{EX}(M\,|\,data). \qquad (6.11)$$

Here

$$\pi_{EX}(\lambda\,|\,M,data) \sim \mathrm{GA}\left(c+r, \sum_{i=1}^{r} t_{(i)} + d + MT^*\right) \qquad (6.12)$$

and

$$\pi_{EX}(M=m\,|\,data) = K\frac{\theta^{m+r}}{m!\left(\sum_{i=1}^{r} t_{(i)} + d + mT^*\right)^{a+r}}; \quad m = 0,1,\ldots, \qquad (6.13)$$

where

$$K^{-1} = \sum_{M=0}^{\infty} \frac{\theta^{m+r}}{m!\left(\sum_{i=1}^{r} t_{(i)} + d + mT^*\right)^{a+r}}.$$

For $\theta > 0$, $K < \infty$, all the moments of $\pi_{EX}(M\,|\,data)$ are finite. Since $\{M\,|\,data\}$ is a discrete distribution, a random sample from the PMF (6.13) can be easily generated. Therefore, the generation of samples from (6.9) can be performed as follows. First generate M from the discrete distribution with the PMF (6.13), and for a given $M = m$, λ can be generated from a $\mathrm{GA}\left(a+r, \sum_{i=1}^{r} t_{(i)} + b + mT^*\right)$. Based on the generated samples, Bayes estimates and HPD credible intervals can be easily constructed.

Alternatively, since here the full conditionals have well-known distributions, the Gibbs sampling is more convenient to be used to compute the Bayes estimates and to construct the credible intervals. It can be easily observed that

$$\pi_{EX}(M\,|\,\lambda,data) \sim \mathrm{POI}\left(\theta e^{-\lambda T^*}\right), \qquad (6.14)$$

where $\pi(\lambda|M,data)$ has already been provided in (6.12). The following algorithm can be used for the above purpose.

Algorithm 6.1

- Step 1: Choose λ_0 and m_0, initial values of λ and M, respectively.
- Step 2: For $i = 1,...,B$, generate λ_i and m_i from $\pi(\lambda|M_{i-1},data)$ and $\pi(M|\lambda_{i-1},data)$, respectively.
- Step 3: Choose a suitable burn-in-period B^*, and discard the initial B^* values of λ_i and m_i.
- Step 4: If we denote $g_i = g(m_i, \lambda_i)$, for $i = B^* + 1,...,B$, the Bayes estimate of $g(M, \lambda)$ can be approximated as

$$\widehat{g}(M,\lambda) \approx \frac{1}{B-B^*} \sum_{i=B^*+1}^{B} g_i$$

- Step 5: To construct the $100(1 - \beta)\%$ HPD credible intervals of $g(M, \lambda)$, first-order g_i for $i = B^* + 1,...,B$, as $g_{(B^*+1)} < \cdots < g_{(B)}$, then construct all the $100(1 - \beta)\%$ credible intervals of $g(M, \lambda)$ as

$$\left(g_{(B^*+1)}, g_{(B^*+1+(1-\beta)B)}\right),...,\left(g_{(\beta B)}, g_{(B)}\right).$$

Choose that interval which has the smallest length; see, for example, Chen and Shao (1999).

6.3.2 Weibull GOS Model

The likelihood function (6.1) is as follows:

$$L_{WE}(N,\alpha,\lambda|data) = \frac{N!}{(N-r)!} \alpha^r \lambda^r \prod_{i=1}^{r} t_{(i)}^{\alpha-1} e^{-\lambda\left(\sum_{i=1}^{r} t_{(i)}^{\alpha} + (N-r)(T^*)^{\alpha}\right)}. \quad (6.15)$$

Therefore, the joint posterior density function of N, α, and λ is given by

$$\pi_{WE}(N,\alpha,\lambda|data) \propto \frac{\theta^N}{(N-r)!} \alpha^{a+r-1} e^{-\alpha\left(b-\sum_{i=1}^{r} \ln t_{(i)}\right)} \lambda^{c+r-1} e^{-\lambda\left(\sum_{i=1}^{r} t_{(i)}^{\alpha} + (N-r)(T^*)^{\alpha} + d\right)}. \quad (6.16)$$

Similarly, as before, the joint posterior density of $M = N - r$, α and λ becomes

$$\pi_{WE}(M,\alpha,\lambda|data) \propto \frac{\theta^M}{M!} \alpha^{a+r-1} e^{-\alpha\left(b-\sum_{i=1}^{r} \ln t_{(i)}\right)} \lambda^{c+r-1} e^{-\lambda\left(\sum_{i=1}^{r} t_{(i)}^{\alpha} + M(T^*)^{\alpha} + d\right)}. \quad (6.17)$$

Moreover, under SEL function

$$\hat{g}_B(M,\alpha,\lambda) = E(g(M,\alpha,\lambda)) = \frac{\sum_{m=0}^{\infty}\int_0^{\infty}\int_0^{\infty} g(m,\alpha,\lambda)\pi_{WE}(m,\alpha,\lambda\mid data)d\alpha d\lambda}{\sum_{m=0}^{\infty}\int_0^{\infty}\int_0^{\infty} \pi_{WE}(m,\alpha,\lambda\mid data)d\alpha d\lambda},$$
(6.18)

is the Bayes estimate of $g(M, \alpha, \lambda)$. In general, it does not have a compact form and, hence, we use the MCMC technique to evaluate (6.18).

The full conditional distribution of α is as follows:

$$\pi_{WE}(\alpha\mid M,\lambda,data) \propto \alpha^{a+r-1} e^{-\alpha\left(b-\sum_{i=1}^{r}\ln t_{(i)}\right)} e^{-\lambda\left(\sum_{i=1}^{r} t_{(i)}^{\alpha} + M(T^*)^{\alpha}\right)},$$
(6.19)

and we have the following result regarding the shape of (6.19).

Lemma 6.1: The full conditional PDF of α as given in (6.19) is log-concave.

Proof: See in the Appendix. ∎

Random samples can be easily generated from (6.19); see, for example, Devroye (1984) or Kundu (2008), Furthermore, the full conditional distributions of λ and M are as follows:

$$\pi_{WE}(\lambda\mid\alpha,M,data) \sim \text{GA}\left(c+r, \sum_{i=1}^{r} t_{(i)}^{\alpha} + M(T^*)^{\alpha} + d\right)$$
(6.20)

and

$$\pi_{WE}(M\mid\alpha,\lambda,data) \sim \text{POI}\left(\theta e^{-\lambda(T^*)^{\alpha}}\right).$$
(6.21)

Therefore, we will be able to generate random samples from the full conditionals of α, λ, and M. Hence using the Gibbs sampling procedure as in Algorithm 6.1, the Bayes estimates and the credible intervals can be constructed.

6.3.3 Generalized Exponential GOS Model

In this case for $N = r, r + 1,..., \lambda > 0$ and $\alpha > 0$, the likelihood function (6.1) becomes

$$L_{GE}(N,\alpha,\lambda\mid data) = \frac{N!}{(N-r)!}\alpha^r \lambda^r e^{-\lambda\sum_{i=1}^{r} t_{(i)}} \prod_{i=1}^{r}(1-e^{-\lambda t_{(i)}})^{\alpha-1}\left(1-(1-e^{-\lambda T^*})^{\alpha}\right)^{N-r}.$$

Bayesian Inference

The posterior distribution of α, λ and N is as follows:

$$\pi_{GE}(N,\alpha,\lambda \mid data) \propto \frac{\theta^N}{(N-r)!} \alpha^{a+r-1} e^{-b\alpha} e^{(\alpha-1)\sum_{i=1}^{r}\ln\left(1-e^{-\lambda t(i)}\right)} \lambda^{c+r-1} e^{-\lambda\left(\sum_{i=1}^{r} t_{(i)}+d\right)}$$

$$\times \left\{1-(1-e^{-\lambda T^*})^\alpha\right\}^{N-r}.$$

The joint posterior distribution of $M = N - r$, α and λ is given by

$$\pi_{GE}(M,\alpha,\lambda\mid data) \propto \frac{\theta^M}{M!} \alpha^{a+r-1} e^{-\alpha\left(b-\sum_{i=1}^{r}\ln(1-e^{\lambda t(i)})\right)} \lambda^{c+r-1} e^{-\lambda\left(d+\sum_{i=1}^{r} t_{(i)}\right)}$$

$$\times e^{-\sum_{i=1}^{r}\ln(1-e^{-\lambda t(i)})} \left\{1-(1-e^{-\lambda T^*})^\alpha\right\}^M. \qquad (6.22)$$

The Bayes estimate of any function $g(M, \alpha, \lambda)$ under the SEL function cannot be obtained in explicit form. Importance sampling can be used for constructing the Bayes estimate and the credible interval of $g(M, \alpha, \lambda)$. After some calculations, it can be seen that

$$\pi_{GE}(M,\alpha,\lambda\mid data) \propto \pi_{GE}(M\mid\alpha,\lambda,data) \times \pi_{GE}(\alpha\mid\lambda,data) \times \pi_{GE}(\lambda\mid data)$$

$$\times h(m,\alpha,\lambda,data), \qquad (6.23)$$

where

$$M \mid \alpha,\lambda,data \sim \text{POI}\left(\theta\left(1-\left(1-e^{-\lambda T^*}\right)^\alpha\right)\right)$$

$$\alpha \mid \lambda,data \sim \text{GA}\left(a+r, b-\sum_{i=1}^{r}\ln\left(1-e^{-\lambda t_{(i)}}\right)\right)$$

$$\lambda \mid data \sim \text{GA}\left(c+r, \sum_{i=1}^{r} t_{(i)}+d\right)$$

and

$$h(m,\alpha,\lambda,data) = \left(e^{\theta\left(1-(1-e^{-\lambda T^*})^\alpha\right)-\sum_{i=1}^{r}\ln\left(1-e^{-\lambda t(i)}\right)}\right) \times \frac{1}{\left(b-\sum_{i=1}^{r}\ln\left(1-e^{-\lambda t_{(i)}}\right)\right)^{a+r}}.$$

It follows that if $b > 0$ and $c > 0$, then the right-hand side of (6.23) is integrable. Moreover, it may be noted that $\left(b-\sum_{i=1}^{r}\ln\left(1-e^{-\lambda t_{(i)}}\right)\right)^{a+r}$ is always positive. We can use the following importance sampling procedure for the purpose.

Algorithm 6.2

- Step 1: Generate λ_1, α_1 and m_1, where

  $\lambda_1 \sim \pi_{GE}(\lambda \mid data)$, $\quad \alpha_1 \sim \pi_{GE}(\alpha \mid \lambda_1, data)$ and $m_1 \sim \pi_{GE}(M \mid \alpha_1, \lambda_1, data)$.

- Step 2: Repeat Step 1 B times and obtain $(m_1, \alpha_1, \lambda_1), \ldots, (m_B, \alpha_B, \lambda_B)$.
- Step 3: Obtain the Bayes estimate of $g(M, \alpha, \lambda)$ as

$$E\big(g(M,\alpha,\lambda)\big) = \frac{\sum_{i=1}^{B} g_i h(m_i, \alpha_i, \lambda_i, data)}{\sum_{i=1}^{B} h(m_i, \alpha_i, \lambda_i, data)}, \quad \text{where } g_i = g(m_i, \alpha_i, \lambda_i).$$

- Step 4: To compute the HPD credible interval of $g(M, \alpha, \lambda)$, first compute

$$w_i = \frac{h(m_i, \alpha_i, \lambda_i, data)}{\sum_{i=1}^{B} h(m_i, \alpha_i, \lambda_i, data)}; \quad i = 1, 2, \ldots, B.$$

  Arrange $\{(g_1, w_1), \ldots, (g_B, w_B)\}$ as $\{(g_{(1)}, w_{[1]}), \ldots, (g_{(B)}, w_{[B]})\}$, where $g_{(1)} < \ldots < g_{(B)}$, and $w_{[i]}$'s are associated with $g_{(i)}$'s.
- Step 5: Then a $100(1-\beta)\%$ credible interval of $g(M, \alpha, \lambda)$ is $(g_{(L\gamma)}, g_{(U\gamma)})$, for $0 < \gamma < \beta$, where

$$\sum_{i=1}^{L_\gamma} w_{[i]} \leq \gamma < \sum_{i=1}^{L_\gamma+1} w_{[i]} \text{ and } \sum_{i=1}^{U_\gamma} w_{[i]} \leq 1+\gamma-\beta < \sum_{i=1}^{U_\gamma+1} w_{[i]}.$$

- Step 6: The $100(1-\beta)\%$ HPD credible interval of $g(M, \alpha, \lambda)$ is $(g_{(L\gamma^*)}, g_{(U\gamma^*)})$, such that the length of the credible interval is minimum; see, for example, Chen and Shao(1999).

6.4 Simulation Experiments

Extensive simulation experiments have been performed for different parameter values and for different termination points (T^*) under all the proposed models. To generate samples from a given distribution function, for a fixed N and for fixed parameter values, first we generate a sample of size N from the given distribution using the simple inverse transformation method.

We order the generated sample, and consider those points which are less than or equal to T^*, and that is the required sample. We have adopted the

Bayesian Inference

Gibbs sampling algorithm as suggested in Section 3 for exponential and Weibull GOS models, and in the case of the generalized exponential GOS model, importance sampling technique has been incorporated. The results have been obtained based on 5000 replications. For Gibbs sampling procedure, for each replication, 2500 iterations are performed, and the initial 500 iterations are taken as burn-in. So, for a particular sample, the Bayes estimate and the credible interval are obtained based on 2000 sample values from the posterior density function. For importance sampling, 2000 sample values have been used for each replication.

The results are presented in Tables 6.1–6.6. For each parameter, we present the average estimates, the average lower and upper limits of the 95%

TABLE 6.1

Exponential GOS Model Results (Fixed Parameters)

		$N = 30, \theta = 30, \lambda = 2, a = 5, b = 2.5$				
Parameters		$T^* = 0.25$	$T^* = 0.5$	$T^* = 0.75$	$T^* = 1.0$	$T^* = 5.0$
r	Observed (average)	11.43	18.87	23.42	26.11	30.01
N	Estimate	30.07	30.19	30.27	30.31	30.05
	HPD	(21.37, 40.23)	(22.63, 39.23)	(24.67, 37.56)	(26.25, 36.24)	(28.23, 31.89)
	(Av. length)	(18.86)	(16.60)	(12.89)	(9.99)	(3.66)
	(Cov. per.)	0.92	0.93	0.93	0.94	0.94
λ	Estimate	2.02	2.10	2.09	2.03	2.03
	HPD	(0.98, 3.18)	(1.06, 3.24)	(1.11, 3.16)	(1.13, 3.09)	(1.41, 2.75)
	(Av. length)	(2.20)	(2.18)	(2.05)	(1.96)	(1.34)
	(Cov. per.)	0.93	0.93	0.94	0.95	0.94

TABLE 6.2

Exponential GOS Model Results (Fixed T^*)

		$T^* = 0.75$				
		$N = 30, \theta = r$	$N = 30 = \theta$	$N = 30 = \theta$	$N = 30 = \theta$	$N = 100 = \theta$
		$\lambda = 2, a = 0$	$\lambda = 3, a = 15$	$\lambda = 2, a = 5$	$\lambda = 2, a = 10$	$\theta = 100, \lambda = 2$
Parameters		$b = 0$	$b = 5$	$b = 2.5$	$b = 5$	$a = 5, b = 2.5$
r	Observed (average)	23.41	26.79	23.16	23.25	77.51
N	Estimate	27.18	30.39	30.25	30.22	100.17
	HPD	(23.47, 32.33)	(26.54, 34.75)	(24.62, 37.55)	(24.54, 37.12)	(87.54, 115.12)
	(Av. length)	(8.86)	(8.21)	(12.93)	(12.58)	(27.58)
	(Cov. per.)	0.93	0.92	0.92	0.93	0.94
λ	Estimate	2.67	3.06	2.03	2.03	2.06
	HPD	(1.27, 4.18)	(1.93, 4.18)	(1.09, 3.18)	(1.21, 2.99)	(1.40, 2.75)
	(Av. length)	(2.91)	(2.25)	(2.09)	(1.78)	(1.36)
	(Cov. per.)	0.94	0.95	0.94	0.94	0.95

TABLE 6.3

Weibull GOS Model Results (Fixed Parameters)

		$N = 30, \theta = 30, \alpha = 2$		$\lambda = 1, a = 5$	$b = 2.5, c = 2, d = 2$	
Parameters		$T^* = 0.5$	$T^* = 1.0$	$T^* = 1.25$	$T^* = 1.5$	$T^* = 2.0$
r	Observed (average)	6.77	18.98	23.19	26.89	29.51
N	Estimate	30.11	29.88	30.19	30.56	30.88
	HPD	(20.78, 40.96)	(21.80, 39.48)	(24.26, 37.89)	(26.56, 36.82)	(29.33, 33.97)
	(Av. length)	(20.19)	(17.67)	(13.63)	(10.26)	(4.64)
	(Cov. per.)	0.91	0.93	0.95	0.94	0.95
α	Estimate	2.03	2.14	2.09	2.15	2.01
	HPD	(1.03, 3.05)	(1.39, 2.91)	(1.36, 2.91)	(1.42, 2.81)	(1.39, 2.68)
	(Av. length)	(2.02)	(1.52)	(1.51)	(1.40)	(1.29)
	(Cov. per.)	0.92	0.92	0.94	0.94	0.95
λ	Estimate	1.11	1.07	1.07	1.03	0.99
	HPD	(0.25, 2.04)	(0.47, 1.88)	(0.48, 1.75)	(0.53, 1.58)	(0.55, 1.45)
	(Av. length)	(1.79)	(1.41)	(1.27)	(1.05)	(0.90)
	(Cov. per.)	0.92	0.94	0.94	0.95	0.95

TABLE 6.4

Weibull GOS Model Results (Fixed T^*)

		$T^* = 1.25$				
		$N = 30, \theta = r$	$N = 30 = \theta$	$N = 30 = \theta$	$N = 30 = \theta$	$N = 100 = \theta$
		$\alpha = 2, \lambda = 1$	$\alpha = 1, \lambda = 1$	$\alpha = 2, \lambda = 1$	$\alpha = 2, \lambda = 1$	$\alpha = 2, \lambda = 1$
		$a = 0, b = 0$	$a = 5, b = 5$	$a = 5, b = 2.5$	$a = 10, b = 5$	$a = 5, b = 2.5$
Parameters		$c = 0, d = 0$	$c = 2, d = 2$	$c = 2, d = 2$	$c = 5, d = 5$	$c = 2, d = 2$
r	Observed (average)	23.67	21.38	23.54	23.76	79.46
N	Estimate	26.99	29.89	30.11	30.18	99.91
	HPD	(23.65, 32.23)	(22.69, 38.76)	(24.19, 37.97)	(24.51, 37.69)	(86.13, 116.24)
	(Av. length)	(8.58)	(16.07)	(13.78)	(13.18)	(30.11)
	(Cov. per.)	0.93	0.93	0.94	0.95	0.95
α	Estimate	2.24	1.09	2.08	2.11	2.05
	HPD	(1.48, 3.21)	(0.69, 1.45)	(1.43, 2.83)	(1.37, 2.71)	(1.63, 2.48)
	(Av. length)	(1.73)	(0.76)	(1.40)	(1.34)	(0.85)
	(Cov. per.)	0.91	0.95	0.94	0.92	0.95
λ	Estimate	1.45	1.09	1.07	1.06	1.03
	HPD	(0.67, 2.27)	(0.45, 1.86)	(0.50, 1.77)	(0.51, 1.62)	(0.66, 1.45)
	(Av. length)	(1.60)	(1.41)	(1.28)	(1.11)	(0.79)
	(Cov. per.)	0.93	0.93	0.94	0.94	0.94

Bayesian Inference

TABLE 6.5

Generalized Exponential GOS Model Results (Fixed Parameters)

	Parameters	$N=30, \theta=30, \alpha=2$		$\lambda=1, a=5$	$b=2.5, c=2, d=2$	
		$T^*=0.5$	$T^*=1.0$	$T^*=1.5$	$T^*=2.0$	$T^*=5.0$
r	Observed (average)	4.39	12.15	18.27	22.11	29.76
N	Estimate	29.24	29.18	29.21	30.12	30.10
	HPD	(19.49, 38.51)	(20.15, 36.28)	(21.55, 37.75)	(23.59, 38.68)	(29.58, 33.10)
	(Av. length)	(18.02)	(16.13)	(16.20)	(15.09)	(3.52)
	(Cov. per.)	0.92	0.92	0.94	0.93	0.93
α	Estimate	2.08	2.10	2.08	2.08	2.02
	HPD	(1.05, 3.29)	(1.17, 3.20)	(1.20, 3.22)	(1.18, 3.22)	(1.22, 2.75)
	(Av. length)	(2.24)	(2.03)	(2.20)	(2.04)	(1.53)
	(Cov. per.)	0.93	0.94	0.95	0.95	0.94
λ	Estimate	1.01	1.07	1.07	1.06	0.97
	HPD	(0.41, 1.92)	(0.59, 1.84)	(0.58, 1.77)	(0.57, 1.67)	(0.60, 1.19)
	(Av. length)	(1.51)	(1.25)	(1.19)	(1.10)	(0.59)
	(Cov. per.)	0.93	0.94	0.94	0.95	0.95

TABLE 6.6

Generalized Exponential GOS Model Results (Fixed T^*)

	Parameters	$T^*=1.25$				
		$N=30, \theta=r$	$N=30=\theta$	$N=30=\theta$	$N=30=\theta$	$N=100=\theta$
		$\alpha=2, \lambda=1$	$\alpha=1, \lambda=1$	$\alpha=2, \lambda=1$	$\alpha=2, \lambda=1$	$\alpha=2, \lambda=1$
		$a=0, b=0$	$a=5, b=5$	$a=5, b=2.5$	$a=10, b=5$	$a=5, b=2.5$
		$c=0, d=0$	$c=2, d=2$	$c=2, d=2$	$c=5, d=5$	$c=2, d=2$
r	Observed (average)	18.18	23.11	18.23	18.17	60.51
N	Estimate	22.57	29.31	29.78	29.88	94.12
	HPD	(18.17, 27.43)	(24.19, 34.37)	(21.54, 37.74)	(21.56, 38.59)	(81.23, 104.45)
	(Av. length)	(9.26)	(10.18)	(16.20)	(16.03)	(23.22)
	(Cov. per.)	0.93	0.91	0.94	0.95	0.94
α	Estimate	2.91	1.03	2.10	2.06	2.05
	HPD	(1.33, 4.44)	(0.67, 1.61)	(1.18, 3.28)	(1.27, 2.94)	(1.53, 2.60)
	(Av. length)	(3.11)	(0.94)	(2.10)	(1.67)	(1.07)
	(Cov. per.)	0.93	0.94	0.94	0.92	0.95
λ	Estimate	1.75	1.12	1.11	1.01	1.04
	HPD	(0.84, 2.56)	(0.77, 2.03)	(0.68, 1.88)	(0.62, 1.62)	(0.88, 1.49)
	(Av. length)	(1.72)	(1.26)	(1.20)	(1.00)	(0.61)
	(Cov. per.)	0.94	0.93	0.94	0.94	0.95

HPD credible intervals, the average lengths and the coverage percentages. For each model, the results are presented in two tables. In the first table, T^* is varied while other parameters are kept fixed, and in the second table, T^* is fixed, where the other parameters are changing. We have considered both informative and noninformative priors. In the case of noninformative priors, we take $a = b = c = d = 0$, and it is improper.

It is observed that in all the cases considered here, the average Bayes estimates are very close to the true parameter values. From Tables 6.1, 6.3, and 6.5, it is observed that as T^* increases the lengths of the HPD credible intervals decrease, as expected. From Tables 6.2, 6.4, and 6.6, it is observed that for fixed N, the performance of the Bayes estimates does not depend much on the hyperparameters. In the case of noninformative priors, the average Bayes estimates of N are slightly smaller than the true N, for all the models, although the length of the average HPD credible intervals is also smaller than the rest. In all the cases considered, it is observed that the coverage percentages are very close to the corresponding nominal value.

Overall, the performance of the proposed methods is quite satisfactory.

6.5 Data Analysis

We use the data set presented in Table 6.7. The data points represent the failure times for errors detected during the development of software and time units are in days. See, for example, Osborne and Severini (2000) for the detailed description of the data. This data set has been analyzed by several authors, for example by Goel and Okumoto (1979), Jelinski and Moranda (1972), Raftery (1987), and Osborne and Severini (2000) using the exponential GOS model. In Table 6.8, we have presented the results compiled by Osborne and Severini (2000) on different point estimates of N, and also two different confidence intervals. Various estimators of N are obtained by taking different stopping time T^*, and using exponential distribution.

We have analyzed the data set using three different GOS distributions namely (i) exponential GOS model (M_1), (ii) Weibull GOS model (M_2),

TABLE 6.7

Data Set

9	21	32	36	43	45	50	58	63	70	71	77
78	87	91	92	95	98	104	105	116	149	156	247
249	250	337	384	396	405	540	798	814	849		

Bayesian Inference

and (iii) generalized exponential GOS model (M_3). Since there is no prior information available, we have assumed improper priors in all these cases. Point estimates of the different parameters are presented in Table 6.9 and 95% HPD credible intervals for different parameters are presented in Tables 6.10 and 6.11.

TABLE 6.8

Estimates of N Compiled by Osborne and Severini (2000)

			Point Estimates			95% Conf. Int.	
Stopping Time (T^*)	r	$\hat{N}_{ML}$	$\hat{N}_{JR}$	$\hat{N}_{1/2}$	$\hat{N}_U$	LR	ILR
50	7	∞	25	17	17	(8, ∞)	(7, 248)
100	18	∞	116	82	65	(30, ∞)	(24, 954)
250	26	31	34	30	33	(26, 94)	(27, 81)
550	31	31	32	31	32	(31, 36)	(31, 38)
800	32	32	32	32	32	(32, 33)	(32, 34)
850	34	34	35	34	34	(34, 37)	(34, 39)

$\hat{N}_{ML}$, maximum likelihood estimate of N; $\hat{N}_{JR}$, estimator proposed by Joe and Reid (1985); $\hat{N}_{1/2}$, integrated likelihood estimator (midpoint prior) by Osborne and Severini (2000); $\hat{N}_U$, integrated likelihood estimator (uniform prior) by Osborne and Severini (2000); LR, likelihood ratio confidence interval; ILR, integrated likelihood ratio confidence interval.

TABLE 6.9

Point Estimates of Parameters for Different Models

		M1		M2			M3		
Stopping Time (T^*)	r	$\hat{N}$	$\hat{\lambda}$	$\hat{N}$	$\hat{\alpha}$	$\hat{\lambda}$	$\hat{N}$	$\hat{\alpha}$	$\hat{\lambda}$
50	7	9.84	0.020	10.25	1.233	0.025	9.17	3.79	0.046
100	18	24.6	0.010	25.87	1.412	0.022	23.50	3.24	0.024
250	26	30.59	0.007	37.25	1.319	0.017	27.38	2.80	0.016
550	31	32.55	0.006	44.55	1.278	0.011	32.02	1.37	0.007
800	32	32.54	0.0056	45.18	1.118	0.009	32.40	1.22	0.006
850	34	35.24	0.0043	48.89	1.178	0.007	35.05	1.02	0.004

TABLE 6.10

95% HPD Credible Intervals of N under Different Models

Stopping Time	r	M1	M2	M3
50	7	(7, 14)	(7, 14)	(7, 14)
100	18	(20, 32)	(21, 33)	(18, 30)
250	26	(26, 36)	(31, 46)	(26, 33)
550	31	(31, 36)	(37, 53)	(31, 37)
800	32	(32, 34)	(38, 55)	(32, 34)
850	34	(34, 48)	(40, 56)	(34, 38)

TABLE 6.11

95% HPD Credible Intervals of Other Parameters under Different Models

Stopping Time (T^*)	r	M1 $\hat{\lambda}$	M2 $\hat{\alpha}$	M2 $\hat{\lambda}$	M3 $\hat{\alpha}$	M3 $\hat{\lambda}$
50	7	(0.004, 0.038)	(0.56, 4.231)	(0.011, 0.068)	(0.79, 8.23)	(0.012, 0.084)
100	18	(0.005, 0.016)	(0.79, 3.891)	(0.009, 0.059)	(1.38, 4.64)	(0.009, 0.033)
250	26	(0.004, 0.012)	(0.82, 3.167)	(0.008, 0.045)	(1.11, 3.68)	(0.006, 0.020)
550	31	(0.003, 0.009)	(0.89, 2.671)	(0.003, 0.021)	(0.76, 2.18)	(0.004, 0.011)
800	32	(0.0035, 0.0079)	(0.91, 2.567)	(0.001, 0.018)	(0.69, 1.96)	(0.003, 0.009)
850	34	(0.0025, 0.0061)	(0.94, 1.789)	(0.001, 0.011)	(0.65, 1.52)	(0.003, 0.007)

Next we consider the selection of an appropriate model out of three models under study. The Bayes factor will be used for this purpose. If we have two models, say, M_i and M_j, then the Bayes factor is given by $2\ln B_{ij}$, where

$$B_{ij} = \frac{Prob(data \mid M_i)}{Prob(data \mid M_j)}.$$

Here $Prob(data|M_i)$ and $Prob(data|M_j)$ are the marginal probabilities of data under M_i and M_j, respectively. To compute $Prob(data|M_i)$, we adopt the following procedure. Observe that

$$Prob(data|M_1) = \sum_{N=0}^{\infty} \int_0^{\infty} P(data, N, \lambda | M_1) d\lambda$$

$$= \sum_{N=r}^{\infty} \int_0^{\infty} P(data | N, \lambda, M_1) P(N, \lambda | M_1) d\lambda$$

$$= \frac{d^c}{\Gamma(c)} \sum_{N=r}^{\infty} \int_0^{\infty} \frac{e^{-\theta}\theta^N}{(N-r)!} \lambda^{r+c-1} e^{-\lambda\left(\sum_{i=1}^{r} t_{(i)} + (N-r)T^* + d\right)}$$

$$= \frac{d^c \theta^r}{\Gamma(c)} \sum_{M=0}^{\infty} \frac{e^{-\theta}\theta^M}{M!} \frac{\Gamma(r+c)}{\left(d + \sum_{i=1}^{r} t_{(i)} + MT^*\right)^{r+c}}.$$

Therefore, for noninformative prior

$$Prob(data \mid M_1) = e^{-r} r^r \Gamma(r) \sum_{M=0}^{\infty} \frac{r^M}{M! \left(\sum_{i=1}^{r} t_{(i)} + MT^*\right)^{r+c}}.$$

For noninformative prior, it can be shown after some calculations that

$$\text{Prob}(data|M_2) = e^{-r} r^r \Gamma(r) \sum_{M=0}^{\infty} \frac{r^M}{M!} \times \int_0^{\infty} \frac{\alpha^{r-1} \Pi_{i=1}^r t_{(i)}^{\alpha-1}}{\left(\sum_{i=1}^r t_{(i)}^\alpha + M(T^*)^\alpha\right)^r} d\alpha$$

$$= e^{-r} r^r \Gamma(r) \sum_{M=0}^{\infty} \frac{r^M}{M!} \times A(M),$$

where

$$A(M) = \int_0^{\infty} \frac{\alpha^{r-1} \prod_{i=1}^r t_{(i)}^{\alpha-1}}{\left(\sum_{i=1}^r t_{(i)}^\alpha + M(T^*)^\alpha\right)^r} d\alpha.$$

Note that $A(M)$ can be easily computed using the importance sampling technique. Similarly, $\text{Prob}(data | M_3)$ can also be obtained using the importance sampling technique.

For the given data set, we have provided the Bayes factors namely $2\ln(B_{12})$, $2\ln(B_{13})$, and $2\ln(B_{23})$ in Table 6.12. From Table 6.12, it is immediate that for all values of T^*, the exponential distribution is a clear choice over the Weibull or generalized exponential distributions. We have also computed the log-predictive likelihood values for the three different models as it has been suggested by Kuo and Yang (1996) for the model choice. We have used the first 26 failures as the training sample and the last 5 failures as the predictive sample. Based on the same noninformative prior, we obtained the log-predictive likelihood values for exponential, Weibull, and generalized exponential as −37.15, −37.94, and −38.13, respectively. Therefore, it also indicates the exponential distribution.

It is clear that a Bayes estimator under noninformative priors behaves very similarly to the different estimators obtained based on the frequentist

TABLE 6.12

Bayes Factor for Model Selection

Stopping Time (T^*)	r	2 log(B_{12})	2 log(B_{13})	2 log(B_{23})
50	7	24.12	12.4	11.7
100	18	57.17	23.8	32.78
250	26	74.67	14.8	59.59
550	31	77.21	31.4	46.61
800	32	80.89	29.7	50.38
850	34	77.19	31.9	45.23

approach. On the other hand, the proposed method has an advantage that it is quite simple to implement, and even for small T^*, it produces estimators which are finite and with finite credible intervals.

6.6 Conclusion

In this chapter, we have considered Bayesian estimation of N and other unknown parameters in GOS models. We have considered three different lifetime distributions namely (i) exponential, (ii) Weibull, and (iii) generalized exponential distributions. Based on fairly general priors, the Bayesian inferences are obtained for the unknown parameters. It may be mentioned that Raftery (1987) considered the same problem and provided a Bayesian solution. But the author has mainly restricted the attention to the exponential family. Although he has indicated the generalization to the non-exponential family, no proper method has been suggested for construction of credible intervals of the unknown parameters. In our case, although we have considered only three models, the method can be extended for other distributions also. Since the implementation of the proposed method is quite straightforward, it can be used quite conveniently in practice in a very general setup.

Acknowledgment

This work is partially supported by the project "Optimization and Reliability Modeling" funded by Indian Statistical Institute.

Appendix

Proof of Lemma 6.1: The full conditional of α is as follows:

$$\pi_{WE}(\alpha \mid M, \lambda, data) = k\alpha^{a+r-1} e^{-\alpha\left(b-\sum \ln t_{(i)}\right)} e^{-\lambda\left(\sum_{i=1}^{r} t_{(i)}^{\alpha} + M(T^*)^{\alpha}\right)},$$

where

$$k^{-1} = \int_0^{\infty} \alpha^{a+r-1} e^{-\alpha\left(b-\sum \ln t_{(i)}\right)} e^{-\lambda\left(\sum_{i=1}^{r} t_{(i)}^{\alpha} + M(T^*)^{\alpha}\right)} d\alpha.$$

Now,

$$\ln\pi(\alpha \mid M, \lambda, data) = \ln k + (\alpha + r - 1)\ln\alpha - \alpha\left(b - \sum \ln t_{(i)}\right) - \lambda g(\alpha),$$

where

$$g(\alpha) = \sum_{i=1}^{r} t_{(i)}^{\alpha} + M(T^*)^{\alpha}.$$

Hence,

$$\frac{d^2 \ln\pi(\alpha \mid M, \lambda, data)}{d\alpha^2} = -\frac{a + r - 1}{\alpha^2} - \lambda \frac{d^2 g(\alpha)}{d\alpha^2}.$$

Now $\dfrac{d^2 g(\alpha)}{d\alpha^2} = \sum_{i=1}^{r}(\ln t_{(i)})^2 t_{(i)}^{\alpha} + M(\ln T^*)^2(T^*)^{\alpha} > 0$ and $a + r - 1 > 0$, so $\dfrac{d^2 \ln\pi(\alpha \mid M, \lambda, data)}{d\alpha^2} < 0$ Hence $\pi_{WE}(\alpha \mid M, \lambda, data)$ is log-concave.

References

Anscombe, F.J. (1961). Estimating a mixed exponential response law. *Journal of the American Statistical Association*, vol. 56, 493–502.

Blumenthal, S. and Marcus, R. (1975). Estimating population size with exponential failure. *Journal of the American Statistical Association*, vol. 70, 913–922.

Chen, M.H. and Shao, Q.M. (1999). Monte Carlo estimation of Bayesian credible and HPD intervals. *Journal of Computational and Graphical Statistics*, vol. 8, 69–92.

DasGupta, A. and Herman, R. (2005). Estimation of binomial parameters when both n and p are unknown. *Journal of Statistical Planning and Inference*, vol. 130, 391–404.

Devroye, L. (1984). A simple algorithm for generating random variables with a log-concave density. *Computing*, vol. 33, 246–257.

Forman, E.H. and Singpurwalla, N.D. (1977). An empirical stopping rule for debugging and testing computer software. *Journal of the American Statistical Association*, vol. 72, 750–757.

Goel, A.L. and Okumoto, K. (1979). Time-dependent error detection model for software reliability and other performance measures. *IEEE Transactions on Reliability*, vol. 28, 206–211.

Gupta, R.D. and Kundu, D. (1999). Generalized exponential distributions. *Australian and New Zealand Journal of Statistics*, vol. 41, 173–188.

Hoel, D.G. (1968), Sequential testing of sample size. *Technometrics*, vol. 10, 331–341.

Jelinski, Z. and Moranda, P.B. (1972). Software reliability research. *Statistical Computer Performance Evaluation*, ed. W. Freiberger, London, Academic Press, 465–484.

Jewell, W.S. (1985). Bayesian extensions to basic model of software reliability. *IEEE Transactions on Software Reliability*, vol. 12, 1465–1471.

Joe, H. (1989), Statistical inference for general order statistics and NHPP software reliability models. *IEEE Transactions on Software Reliability*, vol. 16, 1485–1491.

Joe, H. and Reid, N. (1985). Estimating the number of faults in a system. *Journal of the American Statistical Association*, vol. 80, 222–226.

Johnson, N.L. (1962). Estimation of sample size. *Technometrics*, vol. 4, 59–67.

Kundu, D. (2008). Bayesian inference and life testing plan for the Weibull distribution in presence of progressive censoring. *Technometrics*, vol. 50, 144–154.

Kuo, L. and Yang, T. (1995). Bayesian computation of software reliability. *Journal of Computational and Graphical Statistics*, vol. 4, 65–82.

Kuo, L. and Yang, T. (1996). Bayesian computation for nonhomogeneous Poisson processes in software reliability. *Journal of the American Statistical Association*, vol. 91, 763–773.

Meinhold, R.J. and Singpurwalla, N.D. (1983). Bayesian analysis of common used model for describing software failures. *The Statistician*, vol. 32, 168–173.

Nadarajah, S. (2011). The exponentiated exponential distribution: A survey. *Advances in Statistical Analysis*, vol. 95, 219–251.

Osborne, J.A. and Severini, T.A. (2000). Inference for exponential order statistic models based on an integrated likelihood function. *Journal of the American Statistical Association*, vol. 95, 1220–1228.

Raftery, A.E. (1987). Inference and prediction of a general order statistic model with unknown population size. *Journal of the American Statistical Association*, vol. 82, 1163–1168.

7

Large-Scale Reliability-Redundancy Allocation Optimization Problem Using Three Soft Computing Methods

Mohamed Arezki Mellal
M'Hamed Bougara University

Edward J. Williams
University of Michigan

CONTENTS

- 7.1 Introduction 183
- 7.2 System Reliability-Redundancy Allocation Problem 186
- 7.3 Large-Scale System RRAP 187
- 7.4 Solutions Using Soft Computing Methods 188
  - 7.4.1 Genetic Algorithms 189
  - 7.4.2 Particle Swarm Optimization 190
  - 7.4.3 Cuckoo Optimization Algorithm 191
- 7.5 Results and Discussion 192
- 7.6 Conclusion 194
- References 194

7.1 Introduction

The industrial world is subject to various risks, whether traditional (technical, economic, or societal risks) or emerging (informational or psychosocial risks). According to the International Standard Organization 31010: 2009-Risk management—Principles and guidelines, the word "risk" is defined as follows: "effect of uncertainty on objectives." Thus, the definition provided engineers could be modified: "risk is the combination of an event probability and its consequence." Moreover, a competitive industrial plant should have a high level of system reliability. System reliability belongs to the dependability of the plant. Dependability deals with the industrial risks and is based on several basic concepts whose definitions and interpretations are increasingly

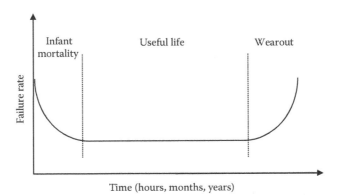

FIGURE 7.1
Bathtub curve.

refined. These concepts are known by the acronym RAMS+C: reliability, availability, maintainability, safety, and cost, but nowadays supplemented with other concepts, such as durability, testability, resilience,... or combinations of these concepts. Dependability reflects the confidence that can be attributed to a system. The International Electrotechnical Commission 60050 (2002) provides some standard definitions for the elements of dependability.

The reliability is the ability of an item E (such as a component part, equipment or system) to perform its required functions under given conditions for a given time interval. It is characterized by the probability denoted $R(t)$ that E is able to perform its functions, under stated conditions, over the time interval $[0, t]$ and knowing that E is functional at time 0. It is defined from the failure rate denoted λ which varies in time, as indicated in Figure 7.1 (called "bathtub curve").

The reliability of a system with a constant failure rate during the useful period is estimated as follows:

$$R(t) = e^{-\lambda t}. \tag{7.1}$$

A failure is the (partial or total) termination of the ability of an item to perform a required function. The failure rate is defined as the number of expected failures per unit in a given time interval (Chowdhury & Koval, 2009; Klyatis, 2011).

The maintainability is the ability of an item E (such as a component part, equipment or system) to undergo maintenance, and hence be returned to a state in which it can perform a required function after a failure, when the maintenance is performed under given conditions and using stated procedures and resources. It is characterized by the probability denoted $M(t)$ that the item E is performing its functions at time t, knowing that E was broken down at the time 0. The maintainability characterizes the promptness of recovering a service expected by the entity after the interruption due to failure.

The availability $A(t)$ is the ability of an item E (such as a component part, equipment or system) to be in a state to perform its required function under given conditions at a given instant of time t or over a given time interval, assuming that the required external resources are provided. The availability is the synthesis of the reliability and the maintainability. It is the proportion of time spent in a state to perform the required functions under given conditions. In the case of a non-repairable entity, it overlaps with the reliability. Besides, for a repairable entity, the availability is appreciated when the failures are rare (better reliability) and shorter (maintainability).

The safety is the ability of an item E (such as a component part, equipment or system) to avoid provoking, under given conditions, critical or catastrophic events which cause damage to people, environment, and property. It is characterized by the probability denoted $S(t)$ that E will not incur, under given conditions, any critical or catastrophic events in time interval $[0, t]$ (Avizienis, Villemeur, & Randell, 2004).

Several metrics of system reliability can be calculated from the probability measures. The following quantities characterize the mean durations (Levin & Kalal, 2003; Pham, 2006):

- MTTF (mean time to failure): It is the average running time of an entity at time 0 to the first failure.
- MUT (mean up time): It is the average time of operation after repair.
- MTTR (mean time to repair): It is the most commonly used term to describe the maintainability of an item. This is the time which separates, for a repairable item, the termination (or degradation) from the recovery of the required functions. It expresses the amount of time required to repair any failures divided by the total number of failures.
- MDT (mean down time): It is the average time of unavailability after failure.
- MTBF (mean time between failures): It is the average time between two consecutive failures (not to be confused with MTTF).
- MTBM (mean time between maintenances): It is a measure of the average time between maintenance (preventive maintenance and repair) for repairable items.

Often, the systems are made up of subsystems and/or components, such as a power system, mechatronic system, or an industrial plant in the general case. These subsystems or components are connected in series or parallel or in a combination configuration (Chowdhury & Koval, 2009). The unit of the frequency failure depends on the operation unit, such as hours, days, months, years, kilometers, tones, solicitations. It should be noted that the choice of the unit (or combination, conversion of the units) promotes the accurate analysis of the behavioral law of the system over a given period. For example, in landing gear of airplanes, it is suitable to use: the distance traveled by the pneumatic system (the wheels) on the tarmac (during takeoff and landing) and the number of landing for the mechanical subsystem. When the system reliability is increased, the frequency and severity of the failure are decreased. It allows us to avoid the industrial risk mentioned above.

The aim of this chapter is to investigate a large-scale system reliability-redundancy allocation problem (RRAP) involving 20 subsystems using genetic algorithms (GAs), particle swarm optimization (PSO), and the cuckoo optimization algorithm (COA). It is to show how the above-mentioned three soft computing methods tackle the problem. The remainder of the chapter is organized as follows: Section 7.2 presents an overview of the system RRAP. A large-scale numerical case study is presented in Section 7.3. The applied soft computing methods are presented in Section 7.4. The obtained results with a discussion are given in Section 7.5. Finally, the last section concludes the chapter.

7.2 System Reliability-Redundancy Allocation Problem

Three methods may be followed to improve the system reliability, namely by increasing the component reliability (called reliability allocation), using redundant components added in parallel (called redundancy allocation), and mixture (called reliability-redundancy allocation). It is an optimization problem subject to the design constraints, such as the volume, weight, and cost. The RRAP is one of the most complex problems in reliability engineering. The goal is to increase the overall system reliability by increasing the reliability of the subsystems and adding redundant components. Therefore, the problem is mixed, as the reliability of the system/each subsystem should be within the interval [0.5, 1] and the number of redundant components within the interval [1, n_{max}].

Several classical mathematical approaches have been used to solve the system reliability optimization problems, such as exact methods (Djerdjour & Rekab, 2001; Hikita, Nakagawa, Nakashima, & Narihisa, 1992; Kulshrestha & Gupta, 1973) and approximate methods (Kolesar, 1967; Ramirez-Marquez, Coit, & Konak, 2004).

Most often, the researchers use soft computing methods to optimize the system reliability. In Garg (2015b), the reliability-redundancy allocation of five subsystems connected in series has been investigated using an efficient biogeography-based optimization algorithm, whereas in Garg (2015a) and Yeh & Hsieh (2011) the cuckoo search (CS) and the artificial bee colony (ABC), respectively, for the same configuration were used. A series-parallel connection has been considered using a modified artificial immune (AI) algorithm (Hsieh & You, 2011), differential evolution (DE) (Liu & Qin, 2015), improved novel global harmony search (INGHS) (Chen, 2006; Ouyang, Gao, Li, & Kong, 2015), modified imperialist competitive algorithm (AR-ICA) (Afonso, Mariani, & Dos Santos Coelho, 2013), CS (Garg, 2015a; Valia, 2014; Valian, Tavakoli, Mohanna, & Haghi, 2013), hybrid cuckoo search and genetic algorithm (CS-GA) (Kanagaraj, Ponnambalam, & Jawahar, 2013), simplified swarm optimization (SSO) (Huang, 2015), and novel artificial fish swarm algorithm (NAFSA) (He, Hu, Ren, & Zhang, 2015). The overall system reliability of an overspeed protection in a gas turbine comprising four subsystems has been investigated using various methods, such as the ABC (Garg, Rani, & Sharma, 2013), a penalty-guided stochastic fractal search (Mellal & Zio, 2016), PSO (Coelho, 2009), and AI algorithm (Hsieh & You, 2011). Some other systems' configuration have been investigated, namely the complex bridge network, life-support system in a space capsule, M-unit structure system, and pharmaceutical plant (Garg & Sharma, 2013; Mellal & Zio, 2016; Agarwal & Sharma, 2010; Garg & Sharma, 2013; Murty & Reddy, 1999; Ravi, Murty, & Reddy, 1997; Zou, Liu, Gao, & Li, 2011). Some works investigate the system reliability optimization as a multi-objective problem (Ardakan & Rezvan, 2018; Ashraf, Muhuri, Lohani, & Nath, 2014; Kumar, Pant, Ram, & Singh, 2017; Muhuri, Ashraf, & Lohani, 2017; Sudeng & Wattanapongsakorn, 2014).

7.3 Large-Scale System RRAP

The system considered here contains 20 subsystems connected in series configuration. The overall system reliability is written as follows (Mellal & Zio, 2016; Zhang, Hu, Shao, Li, & Wang, 2013):

$$\text{Maximize} \quad R_s(r,n) = \prod_{i=1}^{20}\left[1-(1-r_i)^{n_i}\right] \tag{7.2}$$

subject to

$$g_1(r,n) = \sum_{i=1}^{20} v_i n_i^2 \leq V \tag{7.3}$$

$$g_2(r,n) = \sum_{i=1}^{20} \alpha_i (-T/\ln r_i)^{\beta_i} \left[n_i + \exp(n_i/4)\right] \leq C \tag{7.4}$$

$$g_3(r,n) = \sum_{i=1}^{20} w_i n_i \exp(n_i/4) \leq W \tag{7.5}$$

$$0.5 \leq r_i \leq 1, \quad r_i \in [0,1] \subset \mathbb{R}^+; 1 \leq n_i \leq 10, \quad n_i \in \mathbb{Z}^+; i = 1,2,\ldots,20$$

TABLE 7.1

Data of the System

Subsystem i	$10^5\alpha_i$	β_i	v_i	w_i	V	C	W	T (h)
1	0.6	1.5	2	8	500	600	800	1000
2	0.1	1.5	5	9				
3	1.2	1.5	5	6				
4	0.3	1.5	4	10				
5	2.9	1.5	4	8				
6	1.7	1.5	1	9				
7	2.6	1.5	1	9				
8	2.5	1.5	4	7				
9	1.3	1.5	4	9				
10	1.8	1.5	3	8				
11	2.4	1.5	3	9				
12	1.3	1.5	1	8				
13	1.2	1.5	1	7				
14	2.1	1.5	3	10				
15	0.9	1.5	4	6				
16	1.3	1.5	5	7				
17	1.9	1.5	1	7				
18	2.7	1.5	4	8				
19	2.8	1.5	2	9				
20	1.5	1.5	1	9				

where $R_s(r, n)$ is the overall system reliability, r is the vector of component reliabilities for the system, n is the vector of redundancy allocation for the system, $g(\bullet)$ is the set of constraints. V, C, and W are the volume, cost, and weight limits, respectively. α_i and β_i are parameters representing the physical features of each component in the subsystem i, respectively. Table 7.1 reports the data of the system (Mellal & Zio, 2016). The limits considered in this chapter are 500, 600, and 800.

7.4 Solutions Using Soft Computing Methods

Soft computing methods have the ability to provide good results when the human expertise is limited or the problem has a very large range of potential solutions. Bio-inspired soft computing methods are widely used in engineering as they are powerful and allow finding good solutions in a reduced computational time and cost.

7.4.1 Genetic Algorithms

The best-known method is the GA introduced by Holland (Holland, 1975) and is inspired by the human and animal evolution.

The main steps of the GA can be summarized as follows (Chambers, 1999):

- A system of encoding the possible solutions or chromosome structure;
- An initial population of solutions;
- A function to evaluate the solution;
- A method of selecting solutions to be used to produce new solutions;
- Recombination and Mutation operators to create new solutions from those existing;
- Generate the optimal solutions.

Figure 7.2 shows the flowchart of the simple GA (Renner & Ekart, 2003). The GA has several advantages, such as the liability and the parallelism, whereas some of its disadvantages could be listed as follows: it requires a high number of function evaluations and includes many parameters to be handled (Sivanandam & Deepa, 2008a). The GA has been applied to a wide range of engineering problems, such as the design of mechatronic systems (Behbahani & de Silva, 2013), replacement of obsolete components (Mellal, Adjerid, Benazzouz, Berrazouane, & Williams, 2013a,b), multi-objective system reliability with a choice of redundancy strategies (Safari, 2012), cutting parameters (Rai, Brand, Slama, & Xirouchakis, 2011), job-shop

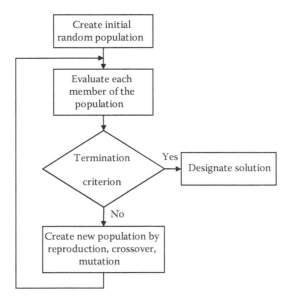

FIGURE 7.2
Flowchart of GA.

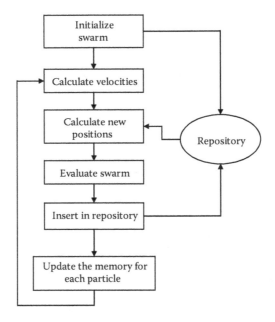

FIGURE 7.3
Flowchart of PSO.

scheduling (Chan, Choy, & Bibhushan, 2011), power distribution system (Torres, Guardado, Rivas-Dávalos, Maximov, & Melgoza, 2013), and signal processing (Han & Chang, 2013).

7.4.2 Particle Swarm Optimization

Another well-known solution method is the PSO (Kennedy & Eberhart, 1995). It is inspired by the moving behavior of some swarms in nature, such as the birds and fishes, where each individual is called a particle. It is based on the position and velocity of each particle.

Its main steps are summarized as follows (Mellal & Williams, 2018; Sivanandam & Deepa, 2008b):

- Initialize the swarm from the solution space.
- Evaluate the fitness of individual particles.
- Modify the particle position, global position, and the velocity.
- Move each particle to a new position.
- Go to step 2, and repeat until convergence or stopping condition is satisfied.

The flowchart of PSO is shown in Figure 7.3 (Sivanandam & Deepa, 2008a). Its main advantage is related to simple calculations and fast execution, whereas the major disadvantages are in fixing the values of its parameters which may diverge the solutions in some cases and it is difficult to implement discrete problems (Mellal & Williams, 2018; Selvi & Umarani, 2010). Various optimization problems have been solved using the PSO, such as machining parameters (Costa, Celano, & Fichera, 2011), electronic circuit design (Vural, Der, & Yildirim, 2011), robotics (Alici, Jagielski, Ahmet Şekercioğlu, & Shirinzadeh, 2006), heat exchanger network synthesis (Pavao, Costa, & Ravagnani, 2017), and supply chain network design (Baris & Ernesto, 2016).

7.4.3 Cuckoo Optimization Algorithm

A prevailing solution method called the COA has been initially developed by Ramin Rajabioun (2011). It is inspired by the lifestyle of the cuckoo bird which lays its eggs in the foreign birds' nests. The cuckoos have the ability to lay eggs that mimic the color and pattern of the eggs of the parasitized birds. However, some eggs are recognized by the host birds and destroyed. On the other hand, when the eggs hatch, some cuckoos' chicks starve as their diet is more than the chicks of the host birds. Therefore, the cuckoos try to find the best living area (Mellal & Williams, 2017; Rajabioun, 2011).

The steps of the COA used in this chapter are as follows (Mellal & Williams, 2016):

- Initialization;
- Egg laying radius;
- Egg recognition;
- Hatching and evaluation;
- Migration (move the mature cuckoo to the new area);
- If the number of cuckoo generations is reached, stop; otherwise, go to the second step.

Figure 7.4 shows the flowchart of the COA applied in this chapter (Mellal & Williams, 2016). The proper convergence is the main advantage of the COA (Mellal & Williams, 2018). The implementation of COA for multiobjective problem is not fluent. The effectiveness of the COA has been proved in solving strong optimization problems, such as machining parameters (Mellal & Williams, 2016; Mellal & Williams, 2015a), lifetime predication (Afzali & Keynia, 2017), replacement of obsolete components (Mellal, Adjerid, & Williams, 2017; Mellal, Adjerid, Williams, & Benazzouz, 2012), electrical power system (Xiao, Shao, Yu, Ma, & Jin, 2017), PID controllers (Rajabioun, 2011), combined heat and power economic dispatch (Mellal & Williams, 2015b), and crack detection (Moezi, Zakeri, & Zare, 2018).

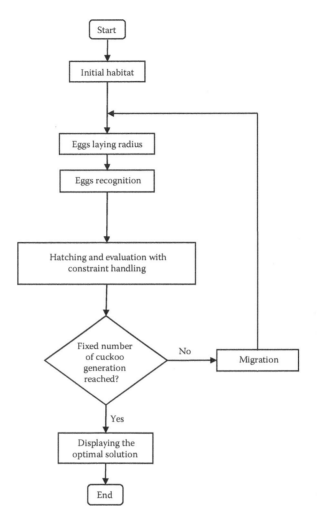

FIGURE 7.4
Flowchart of the applied COA.

7.5 Results and Discussion

The three algorithms have been coded using MATLAB software (2015) and run on a personal computer (G620 2.60 GHz with 4 GB of RAM). The population size (number of individuals or chromosomes), number of particles in the swarm, and number of cuckoos in the area all were fixed at 40 for the GA, PSO, and COA. Table 7.2 reports the obtained results and the required number of function evaluations (NFE) by each algorithm.

Large-Scale RRAP

TABLE 7.2

Optimal Results

Method	$(n_1, n_2, ..., n_{20})$	$(r_1, r_2, ..., r_{20})$	R_s	NFE	CPU (s)
GA	(2, 2, 2, 2, 2, 1, 2, 3, 2, 3, 4, 1, 3, 2, 5, 3, 2, 3, 2, 3)	(0.88681, 0.92059, 0.88549, 0.90911, 0.83291, 0.93366, 0.85113, 0.75681, 0.87442, 0.75788, 0.71254, 0.93403, 0.77369, 0.85899, 0.64968, 0.79825, 0.85595, 0.75353, 0.84706, 0.80085)	0.67517	30,000	512
PSO	(4, 2, 2, 2, 3, 3, 3, 2, 3, 3, 2, 2, 2, 2, 3, 2, 2, 3, 2, 3)	(0.74437, 0.95051, 0.88486, 0.91495, 0.78248, 0.8016, 0.73605, 0.85206, 0.82868, 0.80048, 0.85568, 0.88584, 0.87991, 0.86123, 0.83958, 0.88252, 0.86121, 0.7718, 0.85138, 0.78305)	0.77926	17,000	371
COA	(3, 2, 3, 2, 3, 2, 3, 2, 2, 3, 2, 2, 3, 2, 3, 3, 3, 3, 3, 2)	(0.81429, 0.93583, 0.82047, 0.92784, 0.77002, 0.8753, 0.7821, 0.85952, 0.87833, 0.80336, 0.86, 0.87824, 0.84457, 0.8613, 0.8358, 0.82266, 0.79559, 0.77857, 0.76873, 0.87327)	**0.80377**	**10,000**	229

Bold type represents best value.

From Table 7.2, it can be observed that system reliabilities are 0.67517, 0.77926, and 0.80377 provided by the GA, PSO, and COA, respectively. It clearly shows that the maximum value has been obtained by using the COA ($R_s = 0.80377$). Furthermore, the required numbers of function evaluations by the respective algorithms are 30,000, 17,000, and 10,000. On the other hand, the consumed CPU times by each algorithm are 512 s (GA), 371 s (PSO), and 229 s (COA). Therefore, we can claim that the best method for solving this problem is the COA. Figure 7.5 shows the system reliability improvement.

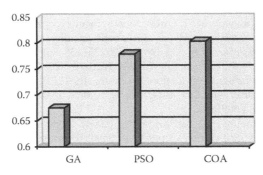

FIGURE 7.5
System reliability provided by each algorithm.

7.6 Conclusion

The goal of this chapter was to present and solve a large-scale system RRAP involving 20 subsystems connected in series. The problem contains 40 decision variables, that is, mixed real-integer optimization problem. The maximizing system reliability was subject to three design constraints, namely the volume, weight, and cost. Three soft computing methods have been implemented, namely the GA, PSO, and the COA. It has been proved that the cuckoo optimization algorithm has increased the overall system reliability and required fewer function evaluations compared to the two other methods. A hybrid approach will be developed to further improve the results. Another point to explore is to investigate the case studied as a multi-objective problem.

References

Afonso, L. D., Mariani, V. C., & Dos Santos Coelho, L. (2013). Modified imperialist competitive algorithm based on attraction and repulsion concepts for reliability-redundancy optimization. *Expert Systems with Applications, 40*(9), 3794–3802. doi:10.1016/j.eswa.2012.12.093.

Afzali, P., & Keynia, F. (2017). Lifetime efficiency index model for optimal maintenance of power substation equipment based on cuckoo optimisation algorithm. *IET Generation, Transmission & Distribution, 11*(11), 2787–2795.

Agarwal, M., & Sharma, V. K. (2010). Ant colony approach to constrained redundancy optimization in binary systems. *Applied Mathematical Modelling, 34*(4), 992–1003. doi:10.1016/j.apm.2009.07.016.

Alici, G., Jagielski, R., Ahmet Şekercioğlu, Y., & Shirinzadeh, B. (2006). Prediction of geometric errors of robot manipulators with Particle Swarm Optimisation method. *Robotics and Autonomous Systems, 54*(12), 956–966. doi:10.1016/j.robot.2006.06.002.

Ardakan, M. A., & Rezvan, M. T. (2018). Multi-objective optimization of reliability–redundancy allocation problem with cold-standby strategy using NSGA-II. *Reliability Engineering and System Safety, 172,* 225–238. doi: 10.1016/j.ress.2017.12.019.

Ashraf, Z., Muhuri, P. K., Lohani, Q. M. D., & Nath, R. (2014). Fuzzy multi-objective reliability-redundancy allocation problem. In *IEEE International Conference on Fuzzy Systems, 6-11 July 2014, Beijing, China* (pp. 2580–2587). doi:10.1109/FUZZ-IEEE.2014.6891889.

Avizienis, A., Villemeur, J. C., & Randell, B. (2004). Dependability and its threats: A taxonomy. In *18th World Computer Congress.* Toulouse, France.

Baris, Y., & Ernesto, M. (2016). Supply chain network design using an enhanced hybrid swarm-based optimization algorithm. In V. N. D. Pandian Vasant, G. W. Weber (Ed.), *Handbook of Research on Modern Optimization Algorithms and Applications in Engineering and Economics* (IGI Global, USA, pp. 95–112).

Behbahani, S., & de Silva, C. W. (2013). Mechatronic design evolution using bond graphs and hybrid genetic algorithm with genetic programming. *IEEE/ASME Transactions on Mechatronics, 18*(1), 190–199.

Chambers, L. D. (1999). *Practical Handbook of Genetic Algorithms: Complex Coding Systems* (CRC Press). Boca Raton, FL, USA.

Chan, F. T. S., Choy, K. L., & Bibhushan. I. (2011). A genetic algorithm-based scheduler for multiproduct parallel machine sheet metal job shop. *Expert Systems with Applications, 38*(7), 8703–8715. doi:10.1016/j.eswa.2011.01.078.

Chen, T. C. (2006). IAs based approach for reliability redundancy allocation problems. *Applied Mathematics and Computation, 182*(2), 1556–1567. doi:10.1016/j.amc.2006.05.044.

Chowdhury, A. A., & Koval, D. O. (2009). *Power Distribution System Reliability: Practical Methods and Applications* (John Wiley). NJ, USA.

Coelho, L. dos S. (2009). An efficient particle swarm approach for mixed-integer programming in reliability-redundancy optimization applications. *Reliability Engineering and System Safety, 94*(4), 830–837. doi:10.1016/j.ress.2008.09.001.

Costa, A., Celano, G., & Fichera, S. (2011). Optimization of multi-pass turning economies through a hybrid particle swarm optimization technique. *The International Journal Of Advanced Manufacturing Technology, 53*, 421–433. doi:10.1007/s00170-010-2861-6.

Djerdjour, M., & Rekab, K. (2001). A branch and bound algorithm for designing reliable systems at a minimum cost. *Applied Mathematics and Computation, 118*(2–3), 247–259. doi:10.1016/S0096–3003(99)00217-9.

Garg, H. (2015a). An approach for solving constrained reliability-redundancy allocation problems using cuckoo search algorithm. *Beni-Suef University Journal of Basic and Applied Sciences, 4*(1), 14–25. doi:10.1016/j.bjbas.2015.02.003.

Garg, H. (2015b). An efficient biogeography based optimization algorithm for solving reliability optimization problems. *Swarm and Evolutionary Computation, 24*, 1–10. doi:10.1016/j.swevo.2015.05.001.

Garg, H., Rani, M., & Sharma, S. P. (2013). An efficient two phase approach for solving reliability–redundancy allocation problem using artificial bee colony technique. *Computers and Operations Research, 40*(12), 2961–2969.

Garg, H., & Sharma, S. P. (2013). Reliability-redundancy allocation problem of pharmaceutical plant. *Journal of Engineering Science and Technology, 8*(2), 190–198.

Han, X., & Chang, X. (2013). An intelligent noise reduction method for chaotic signals based on genetic algorithms and lifting wavelet transforms. *Information Sciences, 218*, 103–118. doi:10.1016/j.ins.2012.06.033.

He, Q., Hu, X., Ren, H., & Zhang, H. (2015). A novel artificial fish swarm algorithm for solving large-scale reliability-redundancy application problem. *ISA Transactions, 59*, 105–113. doi:10.1016/j.isatra.2015.09.015.

Hikita, M., Nakagawa, Y., Nakashima, K., & Narihisa, H. (1992). Reliability optimization of systems by a surrogate-constraints algorithm. *IEEE Transactions on Reliability, 41*(3), 413–480.

Holland, J. H. (1975). *Adaptation in Natural and Artificial Systems* (University of Michigan Press). Ann Arbor, MI, USA. doi:10.1137/1018105.

Hsieh, Y. C., & You, P. S. (2011). An effective immune based two-phase approach for the optimal reliability–redundancy allocation problem. *Applied Mathematics and Computation, 218*(4), 1297–1307.

Huang, C.-L. (2015). A particle-based simplified swarm optimization algorithm for reliability redundancy allocation problems. *Reliability Engineering & System Safety, 142*, 221–230. doi:10.1016/j.ress.2015.06.002.

Kanagaraj, G., Ponnambalam, S. G., & Jawahar, N. (2013). A hybrid cuckoo search and genetic algorithm for reliability-redundancy allocation problems. *Computers and Industrial Engineering, 66*(4), 1115–1124. doi:10.1016/j.cie.2013.08.003.

Kennedy, J., & Eberhart, R. (1995). Particle swarm optimization. *Neural Networks, 4*, 1942–1948. doi:10.1109/ICNN.1995.488968, *1995. Proceedings., IEEE International Conference*.

Klyatis, L. M. (2011). *Accelerated Reliability and Durability Testing Technology* (John Wiley). NJ, USA.

Kolesar, P. J. (1967). Linear programming and the reliability of multi-component systems. *Naval Research Logistics Quarterly, 15*, 317–327.

Kulshrestha, D. K., & Gupta, M. C. (1973). Use of dynamic programming for reliability engineers. *IEEE Transactions on Reliability, R-22*, 240–241.

Kumar, A., Pant, S., Ram, M., & Singh, S. B. (2017). On solving complex reliability optimization problem using multi-objective particle swarm optimization. In *Mathematics Applied to Engineering* (Academic P, Chichester, UK, pp. 115–131).

Levin, M. A., & Kalal, T. T. (2003). Reliability concepts. In *Improving Product Reliability: Strategies and Implementation* (John Wiley, pp. 47–63). Chichester.

Liu, Y., & Qin, G. (2015). A DE algorithm combined with Lévy flight for reliability redundancy allocation problems. *International Journal of Hybrid Information Technology, 8*(5), 113–118.

Mellal, M. A., Adjerid, S., Williams, E. J., & Benazzouz, D. (2012). Optimal replacement policy for obsolete components using cuckoo optimization algorithm based-approach: Dependability context. *Journal of Scientific and Industrial Research, 71*(11), 715–721.

Mellal, M. A., Adjerid, S., Benazzouz, D., Berrazouane, S., & Williams, E. J. (2013a). Obsolescence optimization of electronic and mechatronic components by considering dependability and energy consumption. *Journal of Central South University, 20*(5), 1221–1225. doi:10.1007/s11771-013-1605-9.

Mellal, M. A., Adjerid, S., Benazzouz, D., Berrazouane, S., & Williams, E. J. (2013b). Optimal policy for the replacement of industrial systems subject to technological obsolescence - Using genetic algorithm. *Acta Polytechnica Hungarica, 10*(1), 197–208.

Mellal, M. A., Adjerid, S., & Williams, E. J. (2017). Replacement optimization of industrial components subject to technological obsolescence using artificial intelligence. In 2017 *6th International Conference on Systems and Control, ICSC 2017, Batna, Algeria*. doi:10.1109/ICoSC.2017.7958637.

Mellal, M. A., & Williams, E. J. (2015a). Cuckoo optimization algorithm for unit production cost in multi-pass turning operations. *The International Journal of Advanced Manufacturing Technology, 76*(1–4), 647–656. doi:10.1007/s00170-014-6309-2.

Mellal, M. A., & Williams, E. J. (2015b). Cuckoo optimization algorithm with penalty function for combined heat and power economic dispatch problem. *Energy, 93*, 1711–1718. doi:10.1016/j.energy.2015.10.006.

Mellal, M. A., & Williams, E. J. (2016). Total production time minimization of a multi-pass milling process via cuckoo optimization algorithm. *International Journal of Advanced Manufacturing Technology*. doi:10.1007/s00170-016-8498-3.

Mellal, M. A., & Williams, E. J. (2017). The cuckoo optimization algorithm and its applications. In Samui, P., Roy, S. S., & Balas, V. E (Eds.), In *Handbook of Neural Computation* (Elsevier, pp. 269–277).

Mellal, M. A., & Williams, E. J. (2018). A survey on ant colony optimization, particle swarm optimization, and cuckoo algorithms. Vasant, P., Alparslan-Gok, S. Z., & Weber, G. W (Eds.), In *Handbook of Research on Emergent Applications of Optimization Algorithms* (IGI Global, USA, 37–51). USA.

Mellal, M. A., & Zio, E. (2016). A penalty guided stochastic fractal search approach for system reliability optimization. *Reliability Engineering and System Safety*, 152, 213–227.

Moezi, S. A., Zakeri, E., & Zare, A. (2018). Structural single and multiple crack detection in cantilever beams using a hybrid Cuckoo-Nelder-Mead optimization method. *Mechanical Systems and Signal Processing*, 99. doi:10.1016/j.ymssp.2017.07.013.

Muhuri, P. K., Ashraf, Z., & Lohani, Q. M. D. (2017). Multi-objective Reliability-Redundancy Allocation Problem with Interval Type-2 Fuzzy Uncertainty. *IEEE Transactions on Fuzzy Systems*. doi:10.1109/TFUZZ.2017.2722422.

Murty, B. S. N., & Reddy, P. J. (1999). Reliability optimization of complex systems using modified random-to-pattern search (MRPS) algorithm. *Quality and Reliability Engineering International*, 15(3), 239–243.

Ouyang, H., Gao, L., Li, S., & Kong, X. (2015). Improved novel global harmony search with a new relaxation method for reliability optimization problems. *Information Sciences*, 305, 14–55. doi:10.1016/j.ins.2015.01.020.

Pavao, L. V., Costa, C. B. B., & Ravagnani, M. A. S. S. (2017). Heat Exchanger Network Synthesis without stream splits using parallelized and simplified simulated Annealing and Particle Swarm Optimization. *Chemical Engineering Science*, 158, 96–107. doi:10.1016/j.ces.2016.09.030.

Pham, H. (2006). Basic statistical concepts. In *Springer Handbook of Engineering Statistics* (Springer L, pp. 3–48). London.

Rai, J. K., Brand, D., Slama, M., & Xirouchakis, P. (2011). Optimal selection of cutting parameters in multi-tool milling operations using a genetic algorithm. *International Journal of Production Research*, 49(10), 3045–3068. doi:10.1080/00207540903382873.

Rajabioun, R. (2011). Cuckoo optimization algorithm. *Applied Soft Computing*, 11(8), 5508–5518.

Ramirez-Marquez, J. E., Coit, D. W., & Konak, A. (2004). Redundancy allocation for series-parallel systems using a max-min approach. *IIE Transactions*, 36(9), 891–898.

Ravi, V., Murty, B. S. N., & Reddy, P. J. (1997). Nonequilibrium simulated annealing-algorithm applied to reliability optimization of complex systems. *IEEE Transactions on Reliability*, 46(2), 233–239. doi:10.1109/24.589951.

Renner, G., & Ekart, A. (2003). Genetic algorithms in computer aided design. *Computer-Aided Design*, 35(8), 709–726.

Safari, J. (2012). Multi-objective reliability optimization of series-parallel systems with a choice of redundancy strategies. *Reliability Engineering and System Safety*, 108, 10–20. doi:10.1016/j.ress.2012.06.001.

Sivanandam, S. N., & Deepa, S. N. (2008a). *Introduction to Genetic Algorithms* (Springer B). Berlin.

Sivanandam, S. N., & Deepa, S. N. (2008b). Introduction to particle swarm optimization and ant colony optimization. In *Introduction to Genetic Algorithms* (Springer, Germany, pp. 403–424).

Sudeng, S., & Wattanapongsakorn, N. (2014). A preference-based multi-objective evolutionary algorithm for redundancy allocation problem. In *2014 International Conference on IT Convergence and Security, ICITCS* 2014. doi:10.1109/ICITCS.2014.7021714.

Torres, J., Guardado, J. L., Rivas-Dávalos, F., Maximov, S., & Melgoza, E. (2013). A genetic algorithm based on the edge window decoder technique to optimize power distribution systems reconfiguration. *International Journal of Electrical Power and Energy Systems, 45*(1), 28–34. doi:10.1016/j.ijepes.2012.08.075.

Selvi, V., & Umarani, R. (2010). Comparative analysis of ant colony and particle swarm optimization techniques. *International Journal of Computer Applications, 5*(4), 975–8887. doi:10.5120/908–1286.

Valia, E. (2014). Solving reliability optimization problems by cuckoo search. Yang, X. S (Ed.), In *Cuckoo Search and Firefly Algorithm – Theory and Applications* (Springer I, Switzerland, pp. 195–215).

Valian, E., Tavakoli, S., Mohanna, S., & Haghi, A. (2013). Improved cuckoo search for reliability optimization problems. *Computers and Industrial Engineering, 64*(1), 459–468. doi:10.1016/j.cie.2012.07.011.

Vural, R. A., Der, O., & Yildirim, T. (2011). Investigation of particle swarm optimization for switching characterization of inverter design. *Expert Systems with Applications, 38*(5), 5696–5703. doi:10.1016/j.eswa.2010.10.064.

Xiao, L., Shao, W., Yu, M., Ma, J., & Jin, C. (2017). Research and application of a hybrid wavelet neural network model with the improved cuckoo search algorithm for electrical power system forecasting. *Applied Energy, 198*, 203–222. doi:10.1016/j.apenergy.2017.04.039.

Yeh, W.-C., & Hsieh, T.-J. (2011). Solving reliability redundancy allocation problems using an artificial bee colony algorithm. *Computers & Operations Research, 38*(11), 1465–1473. doi: 10.1016/j.cor.2010.10.028.

Zhang, H., Hu, X., Shao, X., Li, Z., & Wang, Y. (2013). IPSO-based hybrid approaches for reliability–redundancy allocation problems. *Science China Technological Sciences, 56*(11), 2854–2864.

Zou, D., Liu, H., Gao, L., & Li, S. (2011). A novel modified differential evolution algorithm for constrained optimization problems. *Computers and Mathematics with Applications, 61*(6), 1608–1623. doi:10.1016/j.camwa.2011.01.029.

8

A New Distribution-Free Reliability Monitoring Scheme: Advances and Applications in Engineering

Ioannis S. Triantafyllou
University of Thessaly

CONTENTS

8.1 Introduction 199
8.2 The General Setup of the New Monitoring Scheme 201
8.3 Main Characteristics of the Proposed Monitoring Scheme 202
8.4 Numerical Results 207
8.5 An Illustrative Example 209
8.6 Conclusion 212
References 213

8.1 Introduction

Control charts are mainly used for process monitoring in the manufacturing industry. However, their application in fields outside their conventional usage is increasing. It is noticeable that statistical process control can also be proved helpful for reliability monitoring of components or structures. Shewhart-type control charts, especially those used for calculating the number of defects, can be used for monitoring the number of failures per fixed interval.

Generally speaking, failure process monitoring is an important issue for complex or repairable structures. Statistical control charts can be used in this type of failure process monitoring in order to achieve the desirable level of equipment performance. This purpose is often served by plotting the number of failures or breakdowns per time unit. The occurrences of failures contain crucial information about the stability of the system. Especially for highly reliable structures, usage of this information to detect a possible instability is essential. Instability in a system can be seen as a distortion of the distribution of time between failures; the distribution may change or there may be shifts in it. These situations can be adequately monitored through a statistical control chart. For more details about reliability monitoring based on control charts, the interested reader is referred to Xie et al. (2002) or Surucu and Sazak (2009).

Control charting is a statistical technique for detecting possible deviations from target specifications in a production process. Zhang et al. (2006) studied the design of exponential control charts using a sequential sampling scheme to monitor the failure process of a component or structure. Xie et al. (2002) proposed a control scheme based on the cumulative production quantity between observations of two defects to monitor failure processes. They considered the Exponential and Weibull distributions with known parameters for modeling the time between failures to construct control limits from the exact distribution functions. In fact, Xie et al. (2002) studied the use of control charting technique to monitor the failure of components that can be proved appropriate for reliability monitoring.

A great amount of monitoring schemes, already introduced in the literature, is based on the assumption that the process follows a specified probability distribution. However, this argument is not always true in practice and therefore the resulting control charts may not be reliable. To overcome this problem and simultaneously keep the traditional structure of a monitoring scheme, several nonparametric control charts have been proposed in the literature. For example, Balakrishnan et al. (2010) introduced a new distribution-free Shewhart-type control chart based on the location of a single observation and the total number of observations from the test sample that lie between the control limits. Moreover, Triantafyllou (2017) proposed an improved control scheme based on the location of two order statistics of the test sample. Mukherjee and Chakraborti (2012) introduced a control scheme for joint monitoring of location and scale based on the Lepage statistic. In a similar framework, Chowdhury et al. (2014) established a nonparametric control chart exploiting Cucconi statistic. Balakrishnan et al. (2009) proposed the usage of rank-based statistics defined through the test sample observations in order to decide whether the process is in control or not. Some recent advances on the topic can be found in Koutras and Sofikitou (2017) or Balakrishnan et al. (2015). For a detailed presentation of distribution-free monitoring schemes, one may refer to Chakraborti (2014).

In this chapter, we introduce a new distribution-free monitoring scheme based on specific order statistics fetched from a reference sample. The test sample observations that lie between them are taken into account to declare the status of the underlying process. Section 8.2 offers a detailed description of the afore-mentioned family of distribution-free control charts. In Section 8.3, the main characteristics of the members of the above-mentioned class are studied, while in Section 8.4 we carry out numerical experimentation that display the ability of the proposed schemes in detecting shifts of the process distribution. Finally, in Section 8.5 we illustrate the implementation of the new nonparametric control chart to reliability monitoring.

8.2 The General Setup of the New Monitoring Scheme

Let $X_1, X_2, ..., X_m$ denote a reference sample drawn from a process with underlying distribution $F_X(x) = F(x)$. We then specify two-order statistics, say $X_{a:m}, X_{b:m}$, $1 \leq a < b \leq m$. The integers a, b are design parameters and are appropriately selected so that a prespecified level of performance is achieved.

Suppose next that test samples are independently fetched and we focus on clarifying whether the process remains in-control or not. More specifically, let us denote by $Y_1, Y_2, ..., Y_n$ the test sample and by $F_Y(x) = G(x)$ the respective cumulative distribution function. The main goal is to pick up a possible shift in the underlying distribution from $F(x)$ to $G(x)$, that is, to test the null hypothesis $H_0: F(x) = G(x)$ against the two-sided alternative $H_1: F(x) \neq G(x)$.

Within the context described above, we propose the construction of nonparametric control scheme that exploit the location of three test sample observations drawn from the process. More precisely, the ith-, jth-, and the kth-order statistics $Y_{i:n}, Y_{j:n}, Y_{k:n}$, respectively, are selected and used along with the test statistic:

$$R = R(Y_1, Y_2, ..., Y_n; X_{a:m}, X_{b:m}) = |\{t \in \{1, 2, ..., n\} : X_{a:m} \leq Y_t \leq X_{b:m}\}|. \quad (8.1)$$

Note that R is simply the amount of test sample observations that lie between the upper and lower control limit.

The class of distribution-free control charts, introduced in this chapter, makes use of an in-control rule, which embraces the following conditions:

Condition 1. The observations $Y_{i:n}, Y_{j:n}, Y_{k:n}$ of the test sample should lie between the order statistics $X_{a:m}$ and $X_{b:m}$ of the reference sample, namely $X_{a:m} \leq Y_{i:n} \leq Y_{j:n} \leq Y_{k:n} \leq X_{b:m}$.

Condition 2. The number of observations of the Y-sample that lie between the order statistics $X_{a:m}$ and $X_{b:m}$ should be equal to or more than r, namely $R \geq r$.

The conditions stated above define four separate plotting statistics. More precisely, the proposed distribution-free monitoring scheme requires the construction of four different control charts. The first chart, say Chart 1, is based on Condition 1 and depicts the test statistic $Y_{i:n}$. Two additional two-sided charts, Chart 2 and Chart 3 hereafter, are needed for the illustration of observations $Y_{j:n}, Y_{k:n}$ respectively, while a one-sided chart, say, Chart 4, is required for the representation of statistic R. Note that constants a, b, r are design parameters of the proposed monitoring scheme and should be properly determined in order to calculate the corresponding limits of the afore-mentioned control charts. In other words, the observations $X_{a:m}, X_{b:m}$ play the role of the control limits of all two-sided charts, while the parameter r coincides with the control limit of the one-sided Chart 4.

The process is declared to be in-control, if the following conditions hold true:

$$X_{a:m} \leq Y_{i:n}, Y_{j:n}, Y_{k:n} \leq X_{b:m}, R \geq r, \quad i < j < k. \tag{8.2}$$

The large amount of design parameters of the new monitoring scheme, for example m, n, a, b, i, j, k, r, gives the practitioner a notable flexibility for achieving a prespecified level of in-control or out-of-control performance of the resulted control chart. It is quite clear that one may apply the proposed control chart without activating the additional rule referring to the statistic R defined in (8.1). The above argument makes sense, since an appropriate choice of the design parameters can effectively determine the amount of test sample observations that lie between $X_{a:m}, X_{b:m}$ as well. In that case, the modified control chart would produce a signal whenever at least one of the observations $Y_{i:n}, Y_{j:n}$ or $Y_{k:n}$ lies outside of the corresponding control limits. However, the additional parameter r offers the practitioner the opportunity to build the control chart more flexible. Therefore, throughout this chapter we use the parameter r.

8.3 Main Characteristics of the Proposed Monitoring Scheme

Following the general setup for constructing control charts described in detail in Section 8.2, the probability that the proposed monitoring scheme (defined in (8.2)) does not signal is given by

$$p = p(m,n,a,b,i,j,k,r;F,G) = P(X_{a:m} \leq Y_{i:n} \leq Y_{j:n} \leq Y_{k:n} \leq X_{b:m} \text{ and}$$

$$R(Y_1, Y_2, \ldots, Y_n; X_{a:m}, X_{b:m}) \geq r). \tag{8.3}$$

The probability in (8.3) represents the operating characteristic function of the new control scheme. Clearly, the complementary probability corresponds, under the null hypothesis $H_0: F = G$, to the false alarm rate (FAR) of the control scheme as

$$FAR = 1 - p(m,n,a,b,i,j,k,r;F,F).$$

We next deduce a closed formula for the computation of $p = p(m,n,a,b,i,j,k,r;F,G)$ in (8.3). The following proposition offers a more general distributional result which will be proved useful in the sequel.

Proposition 8.1 Let $U_1, U_2, \ldots, U_n$ be a random sample from the uniform distribution in the interval (0, 1) and $U_{i:n}, U_{j:n}, U_{k:n}$ its *i*th-, *j*th-, and *k*th-order statistics, respectively. The following probability:

Reliability Monitoring Scheme

$$q(v,w;r) = P(v \leq U_{i:n} \leq U_{j:n} \leq U_{k:n} \leq w \text{ and } |\{i \in \{1,2,...,n\} : v \leq U_i \leq w\}| \geq r),$$

$$0 \leq v < w \leq 1,$$

can be expressed via

$$q_{c_1,c_2,c_3,c_4}(v,w) = \frac{n!}{(i-c_1-1)!(n-k-c_4)!(c_1+c_2+c_3+c_4+3)!}$$

$$\times v^{i-c_1-1}(w-v)^{c_1+c_2+c_3+c_4+3}(1-w)^{n-k-c_4} \quad (8.4)$$

as follows:

$$q(v,w;r) = \sum_{c_1=0}^{n-3} \sum_{c_4=\max(0,r-c_1-c_2-c_3-3)}^{n-c_1-c_2-c_3-3} q_{c_1,c_2,c_3,c_4}(v,w), \quad 0 \leq v < w \leq 1. \quad (8.5)$$

Proof. The event $\{v \leq U_{i:n} \leq U_{j:n} \leq U_{k:n} \leq w \text{ and } |\{i \in \{1,2,...,n\}: v \leq U_i \leq w\}| \geq r\}$ comes with the scenario that three of the random variables $U_1, U_2, ..., U_n$ achieve values $u_1, u_2, u_3 \in [v,w]$, $u_1 \leq u_2 \leq u_3$ and moreover $c_1 + c_2 + c_3 + c_4 \geq r - 3$ of the remaining uniform variables are within the interval $[v,w]$, while the following ensue:

- $i - c_1 - 1$ of the U_i's are less than or equal to v,
- c_1 of the U_i's are greater than or equal to v and less than u_1,
- c_2 of the U_i's are greater than or equal to u_1 and less than u_2,
- c_3 of the U_i's are greater than or equal to u_2 and less than u_3,
- c_4 of the U_i's are greater than or equal to u_3 and less than w,
- $n - k - c_4$ of the U_i's are greater than or equal to w.

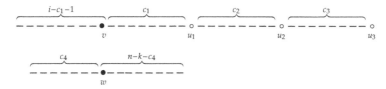

The probability of observing such a scenario can be readily formulated as the multinomial quantity:

$$\frac{n!}{(i-c_1-1)!c_1!c_2!c_3!c_4!(n-k-c_4)!} v^{i-c_1-1}(u_1-v)^{c_1}(u_2-u_1)^{c_2}$$

$$\times (u_3-u_2)^{c_3}(w-u_3)^{c_4}(1-w)^{n-k-c_4}.$$

Upon integrating with respect to u_1, u_2, u_3 for $v \leq u_1 \leq u_2 \leq u_3 \leq w$, we deduce

$$q_{c_1,c_2,c_3,c_4}(v,w) = \int_v^w \int_v^{u_3} \int_v^{u_2} \frac{n!}{(i-c_1-1)!c_1!c_2!c_3!c_4!(n-k-c_4)!} v^{i-c_1-1}(u_1-v)^{c_1}$$
$$\times (u_2-u_1)^{c_2}(u_3-u_2)^{c_3}(w-u_3)^{c_4}(1-w)^{n-k-c_4} \, du_1 \, du_2 \, du_3$$

and therefore the desired result shall be concluded by summing over all possible values of c_1, c_2, c_3, c_4, that is, for all the nonnegative integral values of c_1, c_2, c_3, c_4 satisfying the conditions

$$c_1 + c_2 + c_3 + c_4 + 3 \geq r, \quad c_1 + c_2 + c_3 + c_4 + 3 \leq n,$$

where

$$c_2 = j - i - 1, \quad c_3 = k - j - 1.$$

Applying the transformation $t_1 = u_1 - v/u_2 - v$, we have

$$q_{c_1,c_2,c_3,c_4}(v,w) = B(c_1+1, c_2+1) \int_v^w \int_v^{u_3} \frac{n!}{(i-c_1-1)!c_1!c_2!c_3!c_4!(n-k-c_4)!}$$
$$v^{i-c_1-1} \times (u_2-v)^{c_1+c_2+1}(u_3-u_2)^{c_3}(w-u_3)^{c_4}(1-w)^{n-k-c_4} \, du_2 \, du_3$$

where

$$B(c+1, d+1) = \int_0^1 t^c(1-t)^d \, dt$$

is the well-known Beta function.

We next apply the transformation $t_2 = u_3 - u_2/u_3 - v$ and the following expression is readily deduced:

$$q_{c_1,c_2,c_3,c_4}(v,w) = \frac{n!}{(i-c_1-1)!c_1!c_2!c_3!c_4!(n-k-c_4)!} B(c_1+1, c_2+1)$$
$$\times B(c_3+1, c_1+c_2+2) v^{i-c_1-1}(1-w)^{n-k-c_4}$$
$$\times \int_v^w (u_3-v)^{c_1+c_2+c_3+2}(w-u_3)^{c_4} \, du_3.$$

The proof of Proposition 8.1 is completed by changing the variable u_3 of the last integration to $t_3 = \dfrac{u_3 - v}{w - v}$. ∎

We are now ready to establish a direct way for computing the operating characteristic function (8.3) of the new monitoring scheme. Since

$$p = E_{X_{a:m}, X_{b:m}}[P(X_{a:m} \leq Y_{i:n} \leq Y_{j:n} \leq Y_{k:n} \leq X_{b:m} \text{ and}$$
$$R(Y_1, Y_2, \ldots, Y_n; X_{a:m}, X_{b:m}) \geq r)]$$

Reliability Monitoring Scheme 205

we employ analogous argumentation with the one built up by Balakrishnan et al. (2010) or Triantafyllou (2017) and the operating characteristic function of the new monitoring scheme can be expressed as

$$p = p(m, n, a, b, i, j, k, r; F, G) = \int_0^1 \int_0^t q(GF^{-1}(s), GF^{-1}(t); r) f(s,t) \, ds \, dt, \quad (8.6)$$

where

$$f(s,t) = \frac{m!}{(a-1)!(b-a-1)!(m-b)!} s^{a-1}(t-s)^{b-a-1}(1-t)^{m-b}, \quad 0 < s < t < 1 \quad (8.7)$$

is the joint density function of the order statistics $U_{a:m}$, $U_{b:m}$ of a random sample from the uniform distribution in the interval (0, 1) (see, e.g., Balakrishnan and Ng 2006), while the quantity $q(v, w; r)$ is given in Proposition 8.1. Upon setting $F = G$ in equality (8.6), a closed formula for the calculation of the FAR of the proposed monitoring scheme is deduced.

Proposition 8.2 The *FAR* of the monitoring scheme defined in (8.2) can be computed as follows:

$$FAR = 1 - \sum_{c_1=0}^{n-3} \sum_{c_4=\max(0, r-c_1-c_2-c_3-3)}^{n-c_1-c_2-c_3-3} \frac{\binom{a+i-c_1-2}{a-1}\binom{m-b+n-k+c_4}{m-b}\binom{b+c_1+c_2+c_3+c_4-a+2}{b-a-1}}{\binom{m+n}{n}} \quad (8.8)$$

Proof. When the process is assumed to be in control, equation (8.6) can be written as

$$p = p(m, n, a, b, i, j, k, r; F, F)$$

$$= \int_0^1 \int_0^t q(s, t; r) f(s,t) \, ds \, dt$$

$$= \sum_{c_1=0}^{n-3} \sum_{c_4=\max(0, r-c_1-c_2-c_3-3)}^{n-c_1-c_2-3} \int_0^1 \int_0^t q_{c_1, c_2, c_3, c_4}(s,t) f(s,t) \, ds \, dt.$$

Invoking equation (8.4) and considering that the double integral,

$$\int_0^1 \int_0^t s^{a+i-c_1-2}(t-s)^{b+c_1+c_2+c_3+c_4-a+2}(1-t)^{m+n-b-k-c_4} \, ds \, dt$$

can be rewritten as

$$B(i+b+c_2+c_3+c_4+2, m-b+n-k-c_4+1)$$

$$\times \int_0^1 t^{i+b+c_2+c_3+c_4+1}(1-t)^{m-b+n-k-c_4}\,dt;$$

the outcome we are chasing for is straightforward. ∎

Generally speaking, when a distribution-free monitoring scheme is applied, the signaling events are dependent. This practically means that the average run length (ARL) of the control scheme cannot be computed as the reciprocal of the signaling probability. However, one may exploit the condition-uncondition technique (see, e.g., Balakrishnan et al. 2010) to deduce an expression for the exact run length distribution. More specifically, ARL of a nonparametric chart of the family introduced in this chapter, can be written as

$$ARL = \sum_{k=0}^{\infty} \int_0^1 \int_0^t q^k(G \circ F^{-1}(s), G \circ F^{-1}(t); r)\, f(s,t)\, ds\, dt$$

or equivalently

$$ARL = \int_0^1 \int_0^t \frac{1}{1-q(G \circ F^{-1}(s), G \circ F^{-1}(t); r)} f(s,t)\, ds\, dt. \tag{8.9}$$

It is obvious that for $G = F$, equation (8.9) simplifies as

$$ARL_{in} = \int_0^1 \int_0^t \frac{1}{1-q(s,t;r)} f(s,t)\, ds\, dt, \tag{8.10}$$

where $q(v, v; r)$ is defined in (8.5).

The typical procedure to evaluate the out-of-control performance of a nonparametric control chart requires the out-of-control distribution to be specified. For example, let us consider the case of a process with underlying in-control distribution $N(0, 1)$ and out-of-control distribution $G_\theta = N(\theta, 1)$. Then, the ARL_{out} value takes on the form

$$ARL_{out} = \int_0^1 \int_0^t \frac{1}{1-q(G \circ \Phi^{-1}(s), G \circ \Phi^{-1}(t); r)} f(s,t)\, ds\, dt.$$

The above remark will be proved useful in the sequel for numerical computations.

8.4 Numerical Results

In this section, we bring about several numerical calculations to illustrate the ability of the proposed monitoring scheme to detect possible shift of the underlying distribution. The computations are carried through with the aid of theoretical results presented in Section 8.3.

In Table 8.1, we present the FAR of the proposed control scheme for several designs corresponding to different values of m, n, a, b, i, j, k, r. The calculations were accomplished with the aid of Proposition 8.2. Table 8.1 can be used to design a nonparametric scheme that achieves a prementioned level of in-control performance. The existence of eight different parameters in this monitoring scheme offers the flexibility to fix some of them and then look for the optimal choice of the others. On the other hand, one may look after an adequate design that satisfies our itemized demands. For example, if we carry a reference sample of size $m = 100$ and test samples of size $n = 5$, a FAR almost equal to 0.05 can be accomplished by applying the design $(a, b, i, j, k, r) = (5, 96, 2, 3, 4, 3)$ with exact $FAR = 0.0505$.

In Table 8.2, we present the exact ARL_{in} values for selected designs that meet a prespecified nominal level of in-control performance of the proposed nonparametric monitoring scheme.

For example, if a reference sample of size $m = 200$ is available and test samples of size $n = 25$ are drawn from the process, an in-control ARL almost equal to 370 can be reached by applying the design $(a, b, i, j, k, r) = (24, 158, 8, 9, 10, 7)$ with exact $ARL_{in} = 370.5$.

We next examine the attribution of the new nonparametric monitoring scheme against the ones established by Mukherjee and Chakraborti (2012), Chowdhury et al. (2014) and Triantafyllou (2017) (T-chart hereafter). Tables 8.3 and 8.4 depict the ARL values not only of the proposed control scheme but also of the above-mentioned nonparametric charts under

TABLE 8.1

False Alarm Rates for a Given Design ($b = m - a + 1$)

	Reference Sample Size (m)							
	40		60		100		200	
n	(a, i, j, k, r)	FAR	(a, i, j, k, r)	FAR	(a, i, j, k, r)	FAR	(a, i, j, k, r)	FAR
5	(2, 2, 3, 4, 3)	0.0575	(3, 2, 3, 4, 3)	0.0538	(5, 2, 3, 4, 3)	0.0505	(10, 2, 3, 4, 3)	0.0479
	(3, 2, 3, 4, 3)	0.1090	(4, 2, 3, 4, 3)	0.0870	(6, 2, 3, 4, 4)	0.1145	(15, 2, 3, 4, 3)	0.0988
11	(3, 3, 4, 5, 4)	0.6072	(2, 2, 3, 5, 4)	0.0606	(6, 3, 4, 5, 8)	0.0566	(14, 3, 4, 8, 6)	0.0475
	(4, 3, 4, 6, 4)	0.1074	(3, 2, 3, 5, 4)	0.1107	(5, 3, 4, 5, 9)	0.0972	(19, 3, 4, 8, 6)	0.1009
25	(5, 7, 8, 11, 6)	0.0630	(3, 4, 6, 8, 6)	0.0589	(13, 7, 8, 9, 11)	0.0523	(6, 3, 4, 6, 5)	0.0468
	(6, 7, 8, 11, 6)	0.1088	(4, 4, 6, 8, 6)	0.1090	(16, 7, 8, 9, 6)	0.1113	(9, 3, 4, 5, 7)	0.1098

TABLE 8.2

Designs for a Specific In-Control ARL-Value (ARL_0)

ARL_0	m	n	(LCL, UCL)	(i, j, k, r)	Exact ARL_{in}
370	200	10	(6, 173)	(3, 4, 5, 4)	377.2
		15	(10, 182)	(4, 5, 6, 8)	376.1
		25	(24, 158)	(8, 9, 10, 7)	370.5
	300	10	(10, 280)	(3, 4, 6, 4)	365.9
		15	(20, 253)	(5, 7, 8, 5)	371.6
		25	(27, 229)	(7, 8, 9, 8)	369.1
	400	10	(21, 361)	(4, 5, 6, 4)	375.7
		15	(26, 325)	(6, 7, 8, 5)	367.9
		25	(44, 317)	(8, 9, 10, 9)	369.4
	500	10	(16, 444)	(4, 5, 6, 4)	367.6
		15	(53, 441)	(6, 7, 8, 6)	369.3
		25	(34, 330)	(8, 9, 10, 8)	372.1
500	200	10	(7, 184)	(3, 4, 5, 4)	501.6
		15	(10, 196)	(4, 5, 6, 8)	495.2
		25	(18, 153)	(7, 9, 10, 6)	505.5
	300	10	(10, 290)	(3, 4, 5, 4)	499.6
		15	(18, 253)	(5, 7, 8, 5)	494.0
		25	(26, 240)	(7, 8, 9, 8)	501.7
	400	10	(16, 360)	(4, 5, 6, 4)	498.0
		15	(20, 327)	(6, 7, 8, 5)	500.9
		25	(42, 316)	(8, 9, 10, 9)	498.5
	500	10	(16, 448)	(4, 5, 6, 4)	498.2
		15	(21, 413)	(4, 5, 6, 4)	500.5
		25	(26, 330)	(8, 9, 10, 8)	500.7

normal distribution (θ, δ) and Laplace distribution (θ, δ) respectively. An interesting remark based on the numerical experimentation displayed in Tables 8.3 and 8.4 confirms the ARL-unbiased behavior of the new distribution-free control scheme. Furthermore, the proposed distribution-free control scheme performs better than the other three competitive charts established by Mukherjee and Chakraborti (2012), Chowdhury et al. (2014), and Triantafyllou (2017) for all the cases considered. More specifically, the in-control reference sample in each case is drawn from the corresponding standard distribution with $\theta = 0$ and $\delta = 1$, while several combinations of parameters θ, δ have been examined. When the underlying distribution of the process is assumed to be normal (see Table 8.3), the new control scheme is superior compared to the others for all shifts of the location parameter θ and the scale parameter δ considered. Moreover, Table 8.4 reveals that, under the Laplace distribution, the new monitoring scheme outperforms once again the antagonistic charts.

TABLE 8.3

ARL Values of four Different Control Charts under the $N(\theta, \delta)$ Distribution

θ	δ	New chart	T-chart	Shewhart-Cucconi Chart	Shewhart-Lepage Chart
0	1	465.2	446.6	509.4	513.0
0.25	1	153.3	163.9	253.6	257.6
0.5	1	32.0	51.64	68.6	66.5
1	1	3.3	7.4	7.7	7.7
1.5	1	1.3	2.1	2.1	2.1
2	1	1.1	1.2	1.2	1.2
0	1.25	50.3	61.4	74.5	102.9
0.25	1.25	28.2	35.7	54.9	70.6
0.5	1.25	11.4	17.9	26.2	30.9
1	1.25	2.6	5.0	6.2	6.7
1.5	1.25	1.3	2.1	2.4	2.5
2	1.25	1.1	1.3	1.3	1.4
0	1.5	14.9	20.2	24.3	37.5
0.25	1.5	11.1	15.0	20.4	29.9
0.5	1.5	6.4	9.8	13.4	17.8
1	1.5	2.3	4.1	5.3	6.1
1.5	1.5	1.3	2.1	2.4	2.7
2	1.5	1.1	1.4	1.5	1.6
0	1.75	7.0	10.0	11.7	19.1
0.25	1.75	6.0	8.5	10.7	16.4
0.5	1.75	4.4	6.5	8.4	12.1
1	1.75	2.1	3.5	4.4	5.5
1.5	1.75	1.4	2.1	2.4	2.8
2	1.75	1.1	1.5	1.6	1.8
0	2	4.2	6.2	7.1	11.5
0.25	2	4.0	5.7	6.8	10.8
0.5	2	3.3	4.8	5.8	8.6
1	2	2.0	3.1	3.8	4.8
1.5	2	1.4	2.0	2.4	2.9
2	2	1.1	1.5	1.7	1.9

8.5 An Illustrative Example

For illustration purposes, we shall apply the proposed control scheme for failure process monitoring. More specifically, we use the data given in the work of Xie et al. (2002) (see Table 1 therein). In this particular application, the data correspond to times between failures (measured in hours). The first 20 values comprise historical data that are exponentially distributed with parameter $\lambda = 0.1$. These observations shall be considered as the reference

TABLE 8.4

ARL Values of four Different Control Charts under the *Laplace* (θ, δ) Distribution

θ	δ	New Chart	T-chart	Shewhart-Cucconi Chart	Shewhart-Lepage Chart
0	1	465.2	446.6	509.6	508.3
0.25	1	273.4	276.9	381.6	366.9
0.5	1	124.4	159.2	191.0	159.2
1	1	17.9	45.7	26.5	19.9
1.5	1	3.1	12.2	4.8	4.1
2	1	1.2	3.6	1.8	1.7
0	1.25	87.7	107.5	124.5	153.2
0.25	1.25	60.4	75.9	100.6	121.5
0.5	1.25	34.1	50.4	61.7	66.2
1	1.25	8.33	19.6	14.6	14.0
1.5	1.25	2.4	7.2	4.4	4.2
2	1.25	1.2	2.9	2.0	2.0
0	1.5	30.8	43.1	47.8	66.8
0.25	1.5	23.7	33.2	42.1	55.2
0.5	1.5	15.6	24.3	29.6	36.8
1	1.5	5.4	11.7	10.7	11.1
1.5	1.5	2.1	5.3	4.0	4.1
2	1.5	1.2	2.6	2.1	2.2
0	1.75	15.0	22.8	24.4	36.4
0.25	1.75	12.5	18.7	22.0	32.7
0.5	1.75	9.2	14.7	16.9	23.2
1	1.75	4.1	8.2	7.9	9.2
1.5	1.75	2.0	4.4	3.7	4.0
2	1.75	1.3	2.4	2.1	2.3
0	2	8.9	14.3	14.5	22.9
0.25	2	7.8	12.2	13.6	21.1
0.5	2	6.2	10.0	11.3	16.6
1	2	3.4	6.3	6.3	7.9
1.5	2	1.9	3.8	3.5	3.9
2	2	1.3	2.3	2.1	2.3

sample of size $m = 20$ for the proposed control scheme. The remaining points were simulated when the process mean is assumed to be shifted to 0.01.

The aim is to monitor the above-mentioned failure process by drawing independently from the process, test samples of size $n = 5$. We establish a control scheme (defined in (8.2)) with nominal *FAR* equal to 10%. The prespecified performance of the proposed monitoring scheme can be reached by the design $(a, b, i, j, k, r) = (1, 20, 2, 3, 4, 4)$ with exact *FAR* = 0.0913. Therefore,

Reliability Monitoring Scheme

our interest focuses on the interval created by the 1st and 20th ordered observation of the reference sample. The process will be declared in control if the plotting statistics $Y_{2:5}$, $Y_{3:5}$, $Y_{4:5}$ of the test sample lie between the order statistics $X_{1:20} = 0.47$, $X_{20:20} = 30.02$ and simultaneously at least four observations of each test sample lie between the above control limits.

Figures 8.1–8.4 provide the corresponding charts for all plotting statistics ($Y_{2:5}$, $Y_{3:5}$, $Y_{4:5}$, and R) for all available samples of five observations. It is not difficult to see that, while the Phase I samples are not creating an out of control signal (as expected), the nonparametric scheme signals (the plotting statistics $Y_{3:5}$, $Y_{4:5}$ exceed the upper limit of the corresponding charts, while the statistic R exceed the lower limit) on both test samples in the prospective phase (Phase II) or on fifth and sixth sample overall.

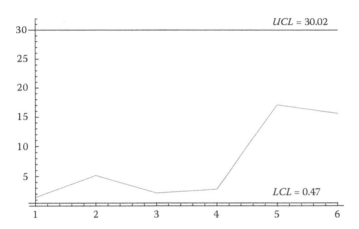

FIGURE 8.1
The plotting statistic $Y_{2:5}$ for failure time data.

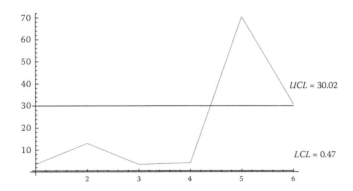

FIGURE 8.2
The plotting statistic $Y_{3:5}$ for failure time data.

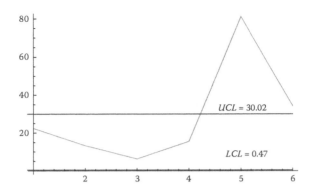

FIGURE 8.3
The plotting statistic $Y_{4:5}$ for failure time data.

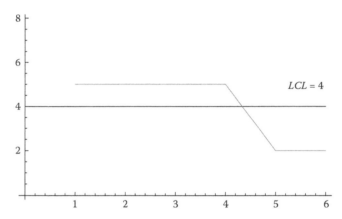

FIGURE 8.4
The plotting statistic r for failure time data.

8.6 Conclusion

This chapter introduces new reliability monitoring scheme based on order statistics. The utility of the proposed control chart arises from the fact that no assumption for the underlying distribution of the process is needed. Moreover, the new control scheme composes a capable statistical tool for failure process monitoring. The numerical outcomes displayed throughout the manuscript, confirm the superiority of the new nonparametric control chart against other antagonistic schemes. To sum up, constructing distribution-free reliability monitoring schemes has already attracted some research attention and it appears that a lot of improvements or modifications can still be realized.

References

Balakrishnan, N. & Ng, H.K.T. (2006). *Precedence-Type Tests and Applications*, Wiley Series in Probability and Statistics, John Wiley & Sons, Hoboken, NJ.

Balakrishnan, N., Paroissin, C. & Turlot, J.-C. (2015). One-sided control charts based on precedence and weighted precedence statistics, *Quality and Reliability Engineering International*, **31**, 113–134.

Balakrishnan, N., Triantafyllou, I.S. & Koutras, M.V. (2009). Nonparametric control charts based on runs and Wilcoxon–type rank–sum statistics, *Journal of Statistical Planning and Inference*, **139**, 3177–3192.

Balakrishnan, N., Triantafyllou, I.S. & Koutras, M.V. (2010). A distribution-free control chart based on order statistics, *Communication in Statistics: Theory & Methods*, **39**, 3652–3677.

Chakraborti, S. (2014). *Nonparametric (Distribution-Free) Quality Control Charts*, Wiley StatsRef: Statistics Reference Online.

Chowdhury, S., Mukherjee, A. & Chakraborti, S. (2014). A new distribution-free control chart for joint monitoring of unknown location and scale parameters of continuous distributions, *Quality and Reliability Engineering International*, **30**, 191–2014.

Koutras, M. V. & Sofikitou, E. (2017). A bivariate semiparametric control chart based on order statistics, *Quality and Reliability Engineering International*, **33**, 183–202.

Mukherjee, A. & Chakraborti, S. (2012). A distribution-free control chart for the joint monitoring of location and scale, *Quality and Reliability Engineering International*, **28**, 335–352.

Surucu, B. & Sazak, H. S. (2009). Monitoring reliability for a three-parameter Weibull distribution, *Reliability Engineering and System Safety*, **94**, 503–508.

Triantafyllou, I.S. (2017). Nonparametric control charts based on order statistics: some advances, *Communication in Statistics: Simulation & Computation*, accepted for publication, doi:10.1080/03610918.2017.1359283.

Xie, M., Goh, T.N. & Ranjan, P. (2002). Some effective control chart procedures for reliability monitoring, *Reliability Engineering and System Safety*, **77**, 143–150.

Zhang, C. W., Xie, M. & Goh, T. N. (2006). Design of exponential control charts using a sequential sampling scheme. *IIE Transactions*, **38**, 1105–1116.

9

Modeling and Simulation of a Sustainable Hybrid Energy System under Changing Power Reliability Index at User End

Anurag Chauhan
Rajkiya Engineering College Banda

CONTENTS

9.1 Introduction	216
9.2 Study Area	217
9.3 Mathematical Modeling	218
9.3.1 Biomass-Operated Generator	218
9.3.2 PV Array	219
9.3.3 Utility Grid	219
9.4 Problem Statement	220
9.4.1 Objective Function	220
9.4.2 Operating Constraints	221
9.4.2.1 Upper and Lower Limit of Power Output	221
9.4.2.2 Grid Sale and Grid Purchase Constraint	222
9.4.2.3 Power Reliability Constraint	222
9.4.2.4 Greenhouse Gas Emission	222
9.5 Optimization Algorithm	222
9.6 Database	223
9.6.1 Hourly Load Demand of the Study Area	223
9.6.2 Hourly Solar Potential	225
9.6.3 Biomass Potential	226
9.6.4 Technical and Economical Data of System Components	226
9.6.5 CO_2 Emission Rate	226
9.7 Results and Discussions	227
9.7.1 Optimum Size of Energy System Model	227
9.7.2 Breakdown of the Total Cost	228
9.7.3 Percentage-Wise Contribution of System Components in Total Cost	228
9.7.4 Electricity Generation of Resources/Grid Sale/Grid Purchase on Seasonal Basis	229
9.8 Conclusions	230
References	230

9.1 Introduction

If the present policy on the use of coal, oil, and gas persist, then by the year 2020, the global temperature is expected to be increased by 2°C. The rise in temperature will result in flooding in lowland areas, increase in the process of desertification, and change in climate all over the world. Therefore, it is required to find the suitable sustainable alternative of fossil fuels. Renewable energy sources require no fossil fuels for power generation and, hence, produce the least negative impact on environment. These sources can be utilized in both utility grid mode and off-grid mode [1–10].

Aktas et al. [11] proposed a novel energy management algorithm for the hybrid energy storage system (HESS) supplied from a three-phase four-wire grid-connected photovoltaic (PV) power system. The considered system comprised battery and ultra-capacitor energy storage units for energy sustainability from the solar-based power generation system. They performed and analyzed eight different operation cases experimentally. They found that the developed algorithm supplied the required load power with the lower operational costs and higher efficiency of the system.

Goel and Sharma [12] presented a comprehensive overview on performance evaluation of a stand-alone, grid-connected, and hybrid renewable energy system for rural areas. They studied several issues of the stand-alone PV system, grid-connected PV system, hybrid energy system, optimization of hybrid system, and plug-in hybrid electric vehicle.

Rajbongshi et al. [13] performed the optimization of a PV/biomass/diesel based and grid-connected hybrid system for rural areas. They considered different load profiles for the size optimization of the system configuration. They found that the cost of energy generation for a grid-connected hybrid system was lower compared to that of an off-grid hybrid system for same pattern of load profiles. Also, they estimated the economic distance limit between grid extension and off-grid system.

Nojavan et al. [14] studied a PV/fuel cell/battery-based hybrid system along with upstream grid to meet out the electrical and thermal load. They have proposed an information gap decision theory (IGDT) technique to model the uncertainty of electrical load. Further, they formulated uncertainty model, robustness function, and opportunity function. Finally, they minimized IGDT-based risk-constrained operation cost of the hybrid system by considering electrical load uncertainty.

Mohamed et al. [15] suggested a particle swarm optimization (PSO) algorithm for the optimal design of hybrid PV-wind energy systems in grid-connected mode. They minimized the total investment cost of the system under the constraints of load-generation balance and loss of load probability as power reliability index. They also considered maximum power point tracking (MPPT) of PV array and wind turbine system. Based on hourly simulation, they found that the cost of supplying the load demand from the hybrid system connected to grid was lower than the cost of energy supplied from the grid only after 25 years.

Sanajaoba and Fernandez [16] investigated the size optimization of three schemes namely PV-battery, wind-battery, and PV-wind-battery system. They minimized the total system cost considering the seasonal changes of load and wind turbine force outage rate. Further, a comparison of results obtained from Cuckoo search algorithm with PSO and genetic algorithm (GA) was performed. Tito et al. [17] accounted socio-demographic load profiles while designing a wind-PV-battery based hybrid system. Based on the analysis, they found that the system cost was affected significantly by the magnitude and temporal positions of the peak demand.

Ahadi et al. [18] minimized the annual capital cost of a hybrid system consisting of wind and PV array resources. They optimized the cost under the constraints of operative reserve of 10% of the load, 50% of wind turbine output, and 25% of PV output. They found that battery bank storage compensated the fluctuations of renewable energy sources. Maleki et al. [19] conducted the size optimization of different combinations of renewable energy sources such as hybrid PV/wind turbine/fuel cell, PV/fuel cell systems and wind turbine/fuel cell. They included the swept area of wind turbine and PV panels and number of storage tanks as decision variables. They minimized the life cycle cost (LCC) by filling the maximum allowable loss of power supply probability.

This chapter presents the modeling and simulation of a sustainable hybrid energy system for different values of power reliability index at user end. Mathematical modeling of the considered system components is discussed in Section 9.2. Further, objective function and constraints are modeled in Section 9.3. Technical and economical data required for this study are given and explained in Section 9.4. The algorithm employed for the optimization of the system is discussed in Section 9.5. Finally, results and discussions for different values of power reliability are summarized in Section 9.6. The main findings of the study are given in Section 9.7.

9.2 Study Area

In this chapter, a small unelectrified village located in the Bijnor district of Indian state of Uttar Pradesh is considered as the study area. The study area is located at the latitude of 29.47°N and longitude of 78.11°E. The population of the village is 421 with a total of 84 households [20].

This village has abundant potential of solar and biomass energy. It receives solar radiation of around 5.14 kWh/m^2/day with more than 300 sunny days. Also, the study area is surrounded by forest. Therefore, a huge amount of biomass in terms of forest foliage and crop residue is available which can be used for the operation of biomass-operated generator. Utilization of these resources in grid-connected environment is recognized as an attractive option to fulfill the domestic energy demand of the village.

9.3 Mathematical Modeling

A configuration of the hybrid system is considered in order to supply energy to the rural households as shown in Figure 9.1. Besides PV array and biomass-operated generator, grid is also incorporated in the system which can supply the deficit load that cannot be met out by the available resources. Also, the surplus electricity can be sold to grid in case available generation exceeds the load demand. Converter is included in order to alter the DC power into AC power.

9.3.1 Biomass-Operated Generator

Power output of a biomass generator depends upon the availability of biomass, calorific value, and operating hours in day. The mathematical model for the power output of the biomass generator is given by following equation:

$$P_B = \frac{Q_B \times CV_B \times \eta_B \times 1000}{365 \times 860 \times H} \qquad (9.1)$$

where Q_B is the availability of forest foliage (tons/year), η_B is the conversion efficiency (20%), CV_B is the calorific value of biomass (kcal/kg), and H is the daily operating hours. The factor of 1/860 is used to convert kcal into kWh.

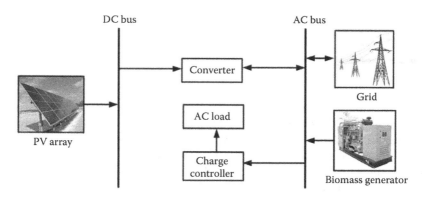

FIGURE 9.1
Schematic of grid connected PV-biomass based hybrid energy system.

9.3.2 PV Array

The power generated from a PV system depends upon different parameters such as incident solar radiation and atmospheric temperature. A mathematical model of PV array is described by following equation:

$$P_{PV}(t) = \left[N_{PV} \times V_{OC}(t) \times I_{SC}(t) \times FF \right]/1000 \tag{9.2}$$

where $P_{PV}(t)$ is the power output of PV array at tth hour, N_{PV} is the number of PV modules, V_{OC} is the open-circuit voltage, I_{SC} is the short-circuit current, and FF is the fill factor.

Short-circuit current and open-circuit voltage at any time 't' can be calculated as

$$I_{SC}(t) = \left[I_{SC,STC} + K_I \{T_C(t) - 25\} \right] \frac{\beta(t)}{1000} \tag{9.3}$$

$$V_{OC}(t) = V_{OC,STC} - K_V T_C(t) \tag{9.4}$$

where $I_{SC,STC}$ is the short-circuit current under STC, K_I is the short-circuit current temperature coefficient, T_C is the cell temperature, β is the hourly average solar radiation (W/m^2), $V_{OC,STC}$ is the open-circuit voltage under STC, and K_V is the open-circuit voltage temperature coefficient.

The cell temperature of the PV module can be calculated as

$$T_C(t) = T_A(t) + \left(\frac{NCOT - 20}{800} \right) \beta(t) \tag{9.5}$$

where NCOT is the nominal cell operating temperature (43°C) and T_A is the ambient temperature.

9.3.3 Utility Grid

In the considered hybrid system, utility grid is incorporated in order to maintain the power reliability at the user end. At hour 't', when available generation is more than the load demand, the surplus amount of energy can be sold to grid which can be mathematically modeled as

$$E_{EE}(t) = \left[(E_{PV}(t) \times \eta_I) + E_{BM}(t) - E_{Load}(t) \right] \tag{9.6}$$

$$E_{GS}(t) = E_{EE}(t) \tag{9.7}$$

where E_{GS} is the amount of electricity sold to grid, E_{EE} is the excess electricity, η_I is the inverter efficiency, E_{PV}, E_{BM}, E_{Load}, respectively, represent hourly PV array generation, biomass gasifier generation and load demand.

When available generation from PV array and biomass generator is not able to fulfill the demand, the remaining deficit demand is supplied by the utility grid as

$$E_{DE}(t) = E_{Load}(t) - \left[(E_{PV}(t) \times \eta_I) + E_{BM}(t)\right] \quad (9.8)$$

$$E_{GP}(t) = E_{DE}(t) \quad (9.9)$$

where E_{DE} is the hourly deficit electricity and E_{GP} is the grid purchase electricity.

9.4 Problem Statement

The techno-economic viability of any project depends on the total cost. Therefore, minimization of total cost (TC) of grid-connected hybrid energy system is considered as the objective function. The total cost has been optimized under technical, social, and environmental constraints.

9.4.1 Objective Function

The total cost of hybrid system is the sum of costs of individual system components during the lifetime of the project and it can be calculated as

$$TC = C_{PV} + C_{BM} + C_{Conv} + C_{grid,\,sale} - C_{grid,\,pur} \quad (9.10)$$

where C_{PV}, C_{BM}, C_{Conv}, respectively, represent the cost of the PV array system, biomass gasifier system, and converter, $C_{grid,\,sale}$ is the total revenue from grid sale, and $C_{grid,\,pur}$ is the total cost of grid purchase.

The total cost of PV array, biomass system, and converter can be calculated by following equations:

$$C_{PV} = \sum_{i=1}^{N_{PV}} (A_i P_i \times CRF) + OM(P_i) + REP(P_i) \quad (9.11)$$

$$C_{BM} = \sum_{j=1}^{N_{BG}} (A_j P_j \times CRF) + OM(P_j) + REP(P_j) \quad (9.12)$$

$$C_{Conv} = \sum_{k=1}^{N_{Conv}} (A_k P_k \times CRF) + OM(P_k) + REP(P_k) \quad (9.13)$$

where CRF is the capital recovery factor, N_{PV}, N_{BG}, and N_{Conv}, respectively, are the numbers PV panels, biogas generators, and converter; A_i, A_j, A_k are the unit cost (INR/kW); P_i, P_j, P_k are the required power capacity (kW); $REP(P_i)$, $REP(P_j)$, $REP(P_k)$ are the replacement cost and $OM(P_i)$, $OM(P_j)$, $OM(P_k)$ are the operation and maintenance (O&M) cost.

The capital recovery factor (CRF) can be determined as

$$CRF = \frac{R_0(1+R_0)^n}{(1+R_0)^n - 1} \tag{9.14}$$

where n is the system lifetime and R_0 is the interest rate.

The excess electricity can be sold to grid to earn revenue which is estimated:

$$C_{grid,\,sale} = \sum_{d=1}^{365} \sum_{t=1}^{24} \left[E_{gs}(d,t) \times c_{gs} \right] \tag{9.15}$$

where $E_{gs}(d, t)$ is the grid sale at hour 't' of day 'd' and c_{gs} is the price of grid sale (INR/kWh).

The deficit amount of electricity can be purchased from the grid and the total cost of grid electricity purchase is calculated as

$$C_{grid,\,pur} = \sum_{d=1}^{365} \sum_{t=1}^{24} \left[E_{gp}(d,t) \times c_{gp} \right] \tag{9.16}$$

where $E_{gp}(d, t)$ is the grid purchase at hour 't' of day 'd', and c_{gp} represent the price of grid purchase (INR/kWh).

The per unit cost of electricity generation (COEG) for the considered system can be estimated as

$$COEG = \frac{TC}{E_D + E_{GS}} \tag{9.17}$$

where E_D is the annual demand (kWh) and E_{GS} is the annual electricity sold to grid (kWh).

9.4.2 Operating Constraints

The total cost of the hybrid system is optimized under the following constraints:

9.4.2.1 Upper and Lower Limit of Power Output

The power output of individual component of the system depends on the number of the units. Therefore, the limits of power output of PV array, biomass generator, and converter are described by the following constraint as:

$$0 \leq N_{PV} \leq N_{PV,\,max} \tag{9.18}$$

$$0 \leq N_{BG} \leq N_{BG,\,max} \tag{9.19}$$

$$0 \leq N_{Conv} \leq N_{Conv,\,max} \tag{9.20}$$

where $N_{PV,\,wmax}$, $N_{BG,\,max}$, and $N_{Conv,\,max}$, respectively, represent the maximum numbers of PV modules, biomass generator unit, and converter unit.

9.4.2.2 Grid Sale and Grid Purchase Constraint

The upper limit of grid sale and grid purchase of electricity in hybrid system are considered that can be expressed as

$$E_{gs}(t) \leq E_{gs,\,max} \tag{9.21}$$

$$E_{gp}(t) \leq E_{gp,\,max} \tag{9.22}$$

where $E_{gs,\,max}$ and $E_{gp,\,max}$ are the upper limit of grid sale and grid purchase of electricity at any hour.

9.4.2.3 Power Reliability Constraint

In this study, the system is designed such that it must fulfill the hourly load demand of the area through the local generation and utility grid. Therefore, unmet load (UL) has been incorporated as the power reliability constraint. It can be determined as

$$UL = \left(\frac{\text{Non-served load for a year}}{\text{Total load for a year}} \right). \tag{9.23}$$

9.4.2.4 Greenhouse Gas Emission

The greenhouse gas emission generated by a hybrid energy system in grid environment has been incorporated as environmental constraint. In this study, emission from the utilization of solar PV, biomass gasifier and grid electricity is considered.

9.5 Optimization Algorithm

In the literature, many metaheuristic algorithms are reported for the size optimization of hybrid energy systems as these algorithms can handle the linear and nonlinear variations of the system components. Metaheuristic algorithms such as GA, PSO, ant colony optimization (ACO), simulated

A Sustainable Hybrid Energy

annealing (SA), harmony search (HS), etc. are extensively used. Among all the algorithms, PSO offers high convergence rate with less time.

Therefore, the PSO algorithm has been used for the optimal design of the considered grid-connected hybrid system. This algorithm is originally discovered by Kennedy and Eberhart in the year 1995. The PSO algorithm searches the global optimum solution vector of a problem based on the concept of social behavior of bird flocking fish schooling, etc.

Stepwise implementation of the PSO algorithm is described as follows [21,22]:

Step 1: The position and velocity of different particles are decided with the help of random variables. The values are generated between the upper and lower limits of the decision vectors.

Step 2: Further, the initial position of individual particle is selected as its pbest. Out of which, gbest has been chosen as the best particle among the total population considered in the algorithm.

Step 3: At each iteration, the velocity of individual particle is modified as

$$v_k^{(t+1)} = \gamma \times \left(w \times v_k^{(t)}\right) + c_1 \times \text{rand}() \times \left(\text{pbest} - x_k^{(t)}\right) + c_2 \times \text{rand}() \times \left(\text{gbest} - x_k^{(t)}\right).$$

Also, the position of individual particle is modified as

$$x_k^{(t+1)} = x_k^{(t)} + v_k^{(t+1)}, \quad k = 1, 2, 3, \ldots, N$$

where rand () is the uniform random values between 0 and 1, t is the iteration index, $\gamma \in [0,1]$, $v_k^{(t)}$ is the velocity of kth particle at generation t, $x_k^{(t)}$ is the current position of kth particle in generation t, w is the inertia weight factor, N is the number of particles in a swarm, c_1 and c_2 are acceleration constants.

Step 4: If the particle crosses the lower and upper limit of allowed range, it is replaced by previous values.

Step 5: Further, the value of objective function is estimated for each particle. At each iteration, pbest and gbest are updated.

Step 6: Finally, simulation is terminated as the stopping criteria is reached.

Based on the steps discussed above, a flowchart of PSO algorithm is prepared, shown in Figure 9.2.

9.6 Database

9.6.1 Hourly Load Demand of the Study Area

Hourly load demand of the study area is depicted in Figure 9.3. Peak demand for season 1, season 2, and season 3 are estimated as 44.57, 36.07, and 22.63 kW,

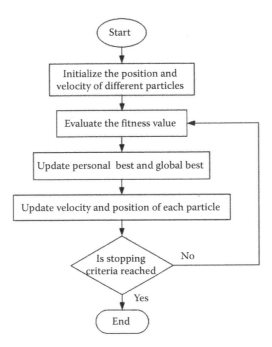

FIGURE 9.2
Flowchart of PSO algorithm.

respectively. The total demand of the area is calculated as 209,295 kWh/year. Among all the seasons, the study area has the highest energy demand during season 1, while the lowest energy demand has been recorded for season 3. Season-wise daily energy demand of the area has been estimated as 755, 640, and 320 kWh for season 1, season 2, and season 3, respectively.

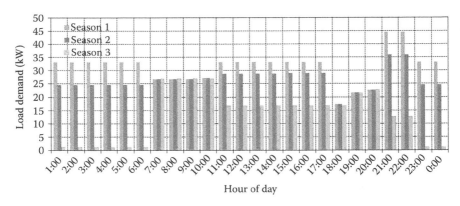

FIGURE 9.3
Hourly load profile of the study area.

9.6.2 Hourly Solar Potential

Hourly solar radiation availability in the study area for different seasons is shown in Figure 9.4. It has been observed that the highest radiation of 800, 700, and 600 W/m² are recorded during season 1, season 2, and season 3, respectively. The Hourly average temperature for the area is shown in Figure 9.5. It has been found that the highest temperature for season 1, season 2, and season 3 are 36.3°C, 33.8°C, and 24°C, respectively. The season-wise average temperature are recorded as 36.3°C, 33.8°C, and 24°C, respectively 27.9°C, 28.4°C and 15.9°C, respectively [23].

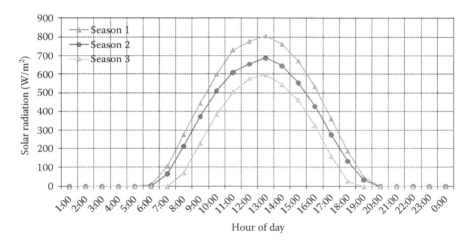

FIGURE 9.4
Hourly average solar radiation availability in the study area [23].

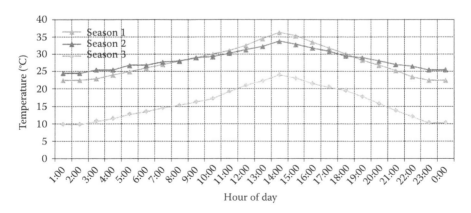

FIGURE 9.5
Hourly average temperature of the study area [23].

9.6.3 Biomass Potential

Biomass potential depends upon the biomass availability in the study area and operating hours. In this study, total biomass availability and operating hours per day are considered as 100 tons/day and 10 h/day. Accordingly, the size of biomass gasifier-based generator has been estimated.

9.6.4 Technical and Economical Data of System Components

Technical and economical data of different system components are given in Tables 9.1–9.3. Economical data include capital cost, replacement cost, and O&M cost of system components. Technical data consist of rating, lifetime, and other specifications of system components. During simulation, 300 W_p PV module, 1 kW size of biomass gasifier system, and 1 kW converter have been considered. Prices of grid sale and grid purchase are taken as INR 6.50 per kWh and INR 3.25 per kWh, respectively, as given in Table 9.4.

Annual real interest rate of 6% and project lifetime of 25 years have been considered in this study.

9.6.5 CO_2 Emission Rate

Emission from the utilization of solar PV, biomass gasifier, and grid purchase for electricity generation are considered and its rates are given in Table 9.5.

TABLE 9.1

Techno-Economical Data of PV Array [24]

S. No.	Indicators	Unit	Value
1	Capital cost	INR	80,000
2	O&M cost	INR	1,600
3	Replacement cost	INR	80,000
4	Rated power output	W_p	300
5	Open-circuit voltage	V	44.80
6	Short-circuit current	A	8.71
7	Short-circuit current temperature coefficient	%	0.0442
8	Open-circuit voltage temperature coefficient	%	−0.2931
9	Fill factor	fraction	0.77
10	Lifetime	year	25

TABLE 9.2

Techno-Economical Data of Biomass Gasifier System [25]

S. No.	Indicators	Unit	Value
1	Capital cost	INR/kW	45,000
2	O&M cost	INR/kW	2,250
3	Replacement cost	INR/kW	45,000
4	Lifetime	year	5

TABLE 9.3
Techno-Economical Data of Converter [26]

S. No.	Indicators	Unit	Value
1	Capital cost	INR/kW	3000
2	O&M cost	INR/kW	0
3	Replacement cost	INR/kW	3000
4	Lifetime	year	10

TABLE 9.4
Price of Grid Sale and Grid Purchase

S. No.	Indicators	Unit	Value
1	Price of grid sale	INR/kWh	6.50
2	Price of grid purchase	INR/kWh	3.25

TABLE 9.5
CO_2 Emission Rate for Different Technologies

S. No.	Energy Technologies	CO_2 Emission (g/kWh)
1	PV array system	130
2	Biomass gasifier system	20
3	Grid electricity	955

As grid electricity is highly dependent on coal-based power plant in the area, the emission rate of grid purchase electricity is the highest among all the considered technologies.

9.7 Results and Discussions

The considered configuration of the hybrid system is optimized using the PSO algorithm for different values of power reliability. Codes are developed in MATLAB environment to obtain the optimum size of system components. Based on hourly simulation, the optimum sizes for different values of power reliability have been reported in Table 9.6.

9.7.1 Optimum Size of Energy System Model

In the hybrid system, both PV array and biomass resources along with utility grid are considered to supply the energy demand of the end user. In simulation, hourly data of load demand, hourly solar radiation, temperature, and biomass generator output are used as input. For 0% UL, the optimum total

TABLE 9.6

Optimum Size of System Components

Power Reliability Value	PPV (kW)	PBMG (kW)	PConv (kW)	TC (in Million INR)	COEG (INR/kWh)
0% UL	48.90	24	40	1.1933	5.28
5% UL	46.50	23	38	1.1362	5.04
10% UL	44.10	22	36	1.0763	4.79
15% UL	41.70	20	34	1.0167	4.56
20% UL	39.30	19	32	0.9568	4.30

cost is calculated as INR 1.1933 million at the COEG of INR 5.28 per kWh. The optimum size of system components are obtained as 48.90 kW PV array, 24 kW biomass gasifier system, and 40 kW converter. The annual electricity from grid purchase and sale are estimated as 55,168 and 16,644 kWh.

Further, the value of power reliability has been changed from 0% to 20% UL. It has been found that the cost of energy of the system varies from INR 5.28 per kWh to INR 4.30 per kWh for the UL changing from 0% to 20%. The optimum sizes of hybrid systems for different power reliability values are reported in Table 9.6.

9.7.2 Breakdown of the Total Cost

Breakdown of the total cost for 0% UL is given in Table 9.7. It has been found that the revenue from grid sale is maximum which accounts for INR 54,092 per year. It has been observed that the considered system needs grid purchase of INR 358,593. The contribution of biomass gasifier in total cost is found to be the highest as INR 476,420 followed by PV array with INR 384,260, grid purchase with INR 358,593, grid sale with INR 54,092, and converter with INR 28,162.

9.7.3 Percentage-Wise Contribution of System Components in Total Cost

Percentage-wise contribution of different components in total cost for 0% UL is depicted in Figure 9.6. It has been found that the contribution of a biomass gasifier system is maximum which accounts for 37% of the total cost followed by PV array system with 29%, grid purchase with 28%, grid sale

TABLE 9.7

Breakdown of Total Cost for 0% Unmet Load

Power Reliability Value	PV Array (INR)	Biomass Gasifier System (INR)	Converter (INR)	Grid Sale (INR)	Grid Purchase (INR)	Total Cost (INR)
0% UL	384,260	476,420	28,162	54,092	358,593	1,193,344

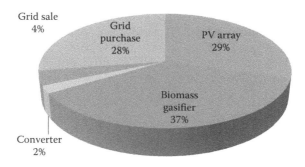

FIGURE 9.6
Percentage-wise contribution of different components in total cost for 0% unmet load.

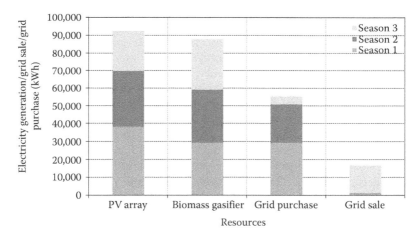

FIGURE 9.7
Season-wise electricity generation of resources/grid sale/grid purchase for 0% unmet load.

with 4%, and converter with 2%. As the replacement cost of biomass gasifier system is the highest, it has the major share in total cost.

9.7.4 Electricity Generation of Resources/Grid Sale/ Grid Purchase on Seasonal Basis

Season-wise generation of different resources/grid sale/grid purchase for the hybrid system is shown in Figure 9.7. It has been observed that a PV array system produces the highest amount of electricity as 37,996 kWh during season 1. While it generates minimum electricity of 22,965 kWh during season 3 due to low availability of solar radiation. The generation from biomass gasifier remains consistent throughout the year. It has been calculated as 29,280, 29,520, and 28,800 kWh for season 1, season 2, and season 3, respectively. The grid sale and grid purchase for each season are also depicted in Figure 9.6.

9.8 Conclusions

This chapter is focused on the modeling and simulation of a grid-connected PV/biomass based hybrid system for an unelectrified village of India. A mathematical model of each system component is presented in detail. Furthermore, the total cost of the system has been formulated by combining capital cost, maintenance cost, replacement cost, grid sale price, and grid purchase price. The total cost has been optimized under the technical, environmental, and power reliability constraints.

Seasonal changes in solar radiation, temperature, and load demand are incorporated in the study. Furthermore, the considered system is optimized for different values of power reliability. It has been found that the cost of generation was reduced with increase in the value of UL. Further, grid sale and grid purchase for different seasons are calculated.

The optimal model consists of 48.90 kW_p PV array system, 24 kW biomass generator, and 40 kW converter. The total cost of this combination has been calculated as INR 1.1933 million at the COEG of INR 5.28 per kWh. The total grid purchase and grid sale for this model are obtained as 55,168 and 16,644 kWh/year, respectively. Therefore, this configuration is recommended for energy access in the area. The results obtained in this study may be helpful for the design and development of hybrid system for other similar unelectrified rural households.

References

1. Rahman MM, Khan MMUH, Ullah MA, Zhang X, Kumar A. (2016) A hybrid renewable energy system for a North American off-grid community. *Energy* 97:151–160.
2. Shams MB, Haji S, Salman A, Abdali H, Alsaffar A. (2016) Time series analysis of Bahrain's first hybrid renewable energy system. *Energy* 103:1–15.
3. Chauhan A, Saini RP. (2016) Techno-economic optimization based approach for energy management of a stand-alone integrated renewable energy system for remote areas of India. *Energy* 94: 138–156.
4. Kanase-Patil AB, Saini RP, Sharma MP. (2011) Development of IREOM model based on seasonally varying load profile for hilly remote areas of Uttarakhand state in India. *Energy* 36: 5690–5702.

5. Chauhan A, Saini RP. (2015) Renewable energy based off-grid rural electrification in Uttarakhand state of India: Technology options, modelling method, barriers and recommendations. *Renewable and Sustainable Energy Reviews* 51:662–681.
6. Chauhan A, Saini RP. (2014) A review on integrated renewable energy system based power generation for stand-alone applications: Configurations, storage options, sizing methodologies and control. *Renewable and Sustainable Energy Reviews* 38:99–120.
7. Bajpai P, Dash V. (2012) Hybrid renewable energy systems for power generation in stand-alone applications: A review. *Renewable and Sustainable Energy Reviews* 16: 2926–36.
8. Chauhan A, Saini RP. (2013) Renewable energy based power generation for stand-alone applications: A review. In: Proceedings of the International Conference on Energy Efficient Technologies for Sustainability (ICEETS) 1:424–8.
9. Kanase-Patil AB, Saini RP, Sharma MP. (2010) Integrated renewable energy systems for off grid rural electrification of remote area. *Renewable Energy* 35: 342–1349.
10. Maleki A, Pourfayaz F, Rosen MA. (2016) A novel framework for optimal design of hybrid renewable energy based autonomous energy systems: A case study for Namin, Iran. *Energy* 98:168–180.
11. Aktas A, Erhan K, Ozdemir S, Ozdemir E. (2017) Experimental investigation of a new smart energy management algorithm for a hybrid energy storage system in smart grid applications. *Electric Power Systems Research* 144:185–196.
12. Goel S, Sharma R. (2017) Performance evaluation of standalone, grid connected and hybrid renewable energy systems for rural application: A comparative review. *Renewable and Sustainable Energy Reviews* 78:1378–1389.
13. Rajbongshi R, Borgohain D, Mahapatra S. (2017) Optimization of PV-biomass-diesel and grid base hybrid energy systems for rural electrification by using HOMER. *Energy* 126:461–474.
14. Nojavan S, Majidi M, Zare K. (2017) Performance improvement of a battery/PV/fuel cell/grid hybrid energy system considering load uncertainty modeling using IGDT. *Energy Conversion and Management* 147:29–39.
15. Mohamed MA, Eltamaly AM, Alolah AI. (2017) Swarm intelligence-based optimization of grid-dependent hybrid renewable energy systems. *Renewable and Sustainable Energy Reviews* 77: 515–524.
16. Sanajaoba S, Fernandez E. (2016) Maiden application of Cuckoo Search algorithm for optimal sizing of a remote hybrid renewable energy system. *Renewable Energy* 96:1–10.
17. Tito SR, Lie TT, Anderson TN. (2016), Optimal sizing of a wind-photovoltaic-battery hybrid renewable energy system considering socio-demographic factors. *Solar Energy* 136:525–532.
18. Ahadi A, Kang S-K, Lee J-H. (2016) A novel approach for optimal combinations of wind, PV, and energy storage system in diesel-free isolated communities. *Applied Energy* 170:101–115.
19. Maleki A, Pourfayaz F, Rosen MA. (2016) A novel framework for optimal design of hybrid renewable energy based autonomous energy systems: A case study for Namin, Iran. *Energy* 98:168–180.
20. Thukral RK, Rahman S (2017) Uttar Pradesh District Factbook Bijnor District. Datanet India Pvt. Ltd. New Delhi.

21. Maleki A, Askarzadeh A. (2014) Comparative study of artificial intelligence techniques for sizing of a hydrogen-based stand-alone photovoltaic/wind hybrid system. *International Journal of Hydrogen Energy* 39:9973–9984.
22. Upadhyay S, Sharma MP. (2015) Development of hybrid energy system with cycle charging strategy using particle swarm optimization for a remote area in India. *Renewable Energy* 77:586–598.
23. NASA. Surface meteorology and solar energy: A renewable energy resource website. Available from: https://eosweb.larc.nasa.gov/sse/ [accessed 7.08.17].
24. PV Module Data. (2017) Data collected from solar modules manufacturers and utilities. Tata Power Solar Systems Catalogue, Noida, India.
25. Biomass Gasifier Data. (2017) Data collected from biomass gasifier manufacturers and utilities. Ankur Scientific Energy Technologies Catalogue, Vadodara, India.
26. Converter Data. (2017) Data collected from battery manufacturers and utilities. Luminous Renewable Energy Solutions Catalogue, Pune, India.

10
Signature Reliability of k-out-of-n Sliding Window System

Akshay Kumar
Tula's Institute, The Engineering and Management College

Mangey Ram
Graphic Era Deemed to be University

CONTENTS

10.1 Introduction 233
10.2 Algorithm for Evaluating the UGF of All the Groups of r
      Consecutive Elements (See Levitin and Dai [9]) 235
10.3 Assessment of m Consecutive Failed Groups to a k-out-of-n SWS 236
10.4 Proposed Algorithms 238
    10.4.1 Algorithm for Evaluating the Reliability of a k-out-of-n
             SWS (See Levitin and Dai [9]) 238
    10.4.2 Algorithm for Calculating the Signature of a
             k-out-of-n SWS with its Reliability Function 238
    10.4.3 Algorithm to Assess the Expected Lifetime of a k-out-of-n
             SWS with Minimum Signature 239
    10.4.4 Algorithm for Evaluating the Expected Value of the
             Component X and Expected Cost Rate of a k-out-of-n SWS
             When Working Elements Are Failed 239
10.5 Illustration 240
    10.5.1 Signature of the k-out-of-n SWS 243
    10.5.2 MTTF of the k-out-of-n SWS 243
    10.5.3 Expected Cost 244
10.6 Conclusion 244
Nomenclature 244
References 245

10.1 Introduction

The sliding window system (SWS) is a generalized form of k-out-of-n:F system which has n linearly ordered multistate elements. Each window can have two states: completely working and totally failed. Application of SWS is found in quality control, service system, manufacturing, radar, and military system. Chiang and Nui [1] discussed the consecutive k-out-of-n:F system in cases when consecutive elements are failed and computed the reliability in lower and upper form. Levitin [2] discussed a linear multistate SWS, which is the generalized form of the consecutive k-out-of-r-from-n:F system, in case of multiple failure and evaluated the reliability of the considered system with the help of universal generating function (UGF). Koucky [3] evaluated the reliability of the k-out-of-n system with failure elements, and concluded that elements not need to be independent and identically distributed (i.i.d.). Levitin [4] considered a linear multistate multiple SWS, which is the generalized form of the linear consecutive k-out-of-r-from-n:F system, in case of multiple failures and computed the system reliability with the help of UGF. Habib et al. [5] discussed the reliability of a linear consecutive k-out-of-r-from-n:G system in case of multistate failure using the total probability theorem. Ram and Singh [6] considered a complex system with common cause failure and where each element could have constant failure rate. They determined the system reliability and cost analysis using the supplementary variable technique. Levitin and Ben-Haim [7] determined the reliability of a consecutive SWS using an algorithm based on the UGF technique; the considered system fails if the sum of the performance rate is lower than the total allocation weight. Levitin and Dai [8] discussed the reliability of linear m-consecutive k-out-of-r-from-n:F systems in case of multiple failure elements. Levitin and Dai [9] considered the k-out-of-n SWS in case of multiple failures and computed the reliability of the proposed system using UGF. Xiang and Levitin [10] generalized the linear multistate SWS which consisted of n linearly multistate windows. They evaluated the reliability of a combined m-consecutive and k-out-of-n SWS using the UGF technique. Ram and Singh [11] discussed the reliability, availability, and cost analysis of two independent repairable subsystems using the supplementary variable technique, Laplace transformation, and Gumbel-Hougaard family copula technique. Ram [12] discussed and reviewed the engineering system and physical science and provided different methods for computing system reliability. Pham [13] discussed the modeling of complex systems both hardware and software and calculated the reliability of the considered system. Negi and Singh [14] studied the non-repairable complex system which had two binary subsystems, namely, weighted A-out-of-G:g and weighted l-out-of-b:g and computed the reliability and sensitivity using UGF. Ram and Davim [15] measured the reliability of multistate systems, optimization of multistate systems, and continuous multistate systems using new computational techniques applied to probabilistic and non-probabilistic safety assessment.

In the context of signatures, Shapley [16] and Owen [17,18] discussed the game theory on the basis of random variable and evaluated the probability by extending the game theory. Samaniego [19] introduced the concept of signatures on the basis of a coherent system. A coherent system can have monotone and its elements relevant to each other. Signatures are widely used in communication networks, reliability economics, etc. Kocher et al. [20] compared the coherent systems when elements having i.i.d. and computed signatures and expected lifetime. Boland and Samaniego [21] described the properties of signature reliability of complex systems and compared the signatures of the different systems. They presented applications of signatures in reliability economics and communication networks. Samaniego [22] examined the properties of a coherent system in terms of signatures. He developed a new approach of coherent system and defined its application in networks and communications. Navarro and Rychlik [23] obtained the reliability function in upper and lower form and computed expected lifetime of a coherent system by using the Samaniego concept of i.i.d elements. Navarro et al. [24] discussed the reliability function based on signatures having i.i.d. elements. They computed the system signature with the help of stochastic ordering properties for the coherent system. Samaniego et al. [25] discussed the system properties on the basis of dynamic signatures with the help of ordered statistical methods. Da et al. [26] evaluated the signature of the considered system and the redundancy system also. They determined the signature of a coherent system which has a large number of elements. Marichal and Mathonet [27] determined the system signature on the basis of reliability function. They used a different formula for evaluating signatures with the help of structure function using i.i.d. components. Coolen [28] discussed the nature of signatures with the help of system structure function and described the definition, properties, and applications of signatures. Kumar and Singh [29–31] determined the signature of sliding window coherent system, k-out-of-n system, and linear multistate SWS having an i.i.d. component and calculated different measures such as signature, expected lifetime, cost, and Barlow-Proschan index.

From the above discussion, it becomes clear that many researchers computed the system reliability of binary and MSS SWS with different methods. We also studied a k-out-of-n SWS with i.i.d. component to evaluate the reliability characteristic such as signature, expected lifetime, and Barlow-Proschan index with the help of Owen's method and UGF technique.

10.2 Algorithm for Evaluating the UGF of All the Groups of r Consecutive Elements (See Levitin and Dai [9])

Step 1. Compute UGF $U_{1-r}(z)$ as follows:

$$U_{1-r}(z) = z^{x_0} \quad (x_0 \text{ consists of } r \text{ zeros}). \tag{10.1}$$

Step 2. Obtain UGF of individual MSE $u(z)$ using $\underset{\leftarrow}{\otimes}$ as follows:

$$u_a(z) \underset{\leftarrow}{\otimes} u_j(z) = \sum_{l=1}^{A_l} q_{a,l} z^{x_{a,l}} \underset{\leftarrow}{\otimes} \sum_{b=1}^{B_l} P_{i,b} z^{g_{i,l}}$$

$$= \sum_{l=1}^{A_l} \sum_{b=1}^{B_l} q_{a,l} P_{i,b} z^{\phi\left(x_{a,l}, g_{i,b}\right)} \qquad (10.2)$$

where ϕ is an arbitrary vector of x and g shift all vector elements one position left.

Step 3. Calculate $U_{i+1-r}(z)$ using operator $\underset{\leftarrow}{\otimes}$ in a sequence as follows:

$$U_{i+1-r}(z) = U_{i-r}(z) \underset{\leftarrow}{\otimes} u_i(z) \quad \text{for } i = 1, 2, \ldots, n.$$

Step 4. Evaluate all possible groups r consecutive of MSE applying operator $\otimes$ as follows:

$$U_i(z) = U_1(z), \ldots, U_{n+1-r}(z).$$

10.3 Assessment of m Consecutive Failed Groups to a k-out-of-n SWS

The UGF of r consecutive groups is given by

$$U_a(z) = \sum_{a=1}^{A_a} q_{a,l} z^{y_{a,l}}.$$

Modifying $U_a(z)$ within an integer counter $C_{a,l}$, we have

$$U_a(z) = \sum_{l=1}^{A_a} q_{a,l} z^{C_{a,l}, y_{a,y}}$$

where m_a = total number of combination of $C_{a,l}$ and $x_{a,l}$.

Now, assign an initial value 0 and modify the equations (10.1) and (10.2):

$$U_{1-r}(z) = z^{0, y_0}$$

$$U_a(z) \otimes u_i(z) = \sum_{l=1}^{A_a} q_{a,l} z^{x_{a,l}, x_{a,l}} \underset{\leftarrow}{\otimes} \sum_{b=1}^{B_i} P_{i,b} z^{g,b}$$

$$U_a(z) = \sum_{l=1}^{A_a} \sum_{b=1}^{B_e} q_{a,l} P_{i,b} z^{\rho\left(C_{a,l},\, \sigma\left(\varphi(y_{i,l}, g_{i,b})\right),\, \phi(y_{a,l}, g_{i,b})\right)}$$

where $\rho(C_g, x) = \begin{cases} C_{g+1} & \text{if } x < w \\ 0 & \text{if } x \geq w \end{cases}$.

UGF of failure probability is expressed as follows:

$$\partial(U_a(z)) = \sum_{l=1}^{m_a} q_{a,l} 1(C_{a,l} = m).$$

The failure probability (E_i) of the system can be computed as the sum of the probabilities of mutually exclusive events as follows:

$$E = E_1 + E_2(1 - E_1) + \cdots + E_{n-r-M+2} \prod_{i=1}^{n-r-M+1} (1 - E_i).$$

To obtain the probability $E_e \prod_{i=1}^{e=1}(1 - E_i)$ one can remove all term with $C_{m+e-1,l} = m$ from $U_{m+e-1}(z)$ to get

$$U_{m+e}(z) = U_{m+e-1}(z) \otimes u_{m+r+e-1}(z).$$

Further, any term l of $U_a(z)$ with the counter value can be obtained as (see Levitin and Dai [9])

$$C_{a,l} < m - n + a + r - 1. \tag{10.3}$$

The system reliability of the sets of B consecutive groups of D consecutive MSE $E_1, E_2, \ldots, E_{g-D+B+2}$ can be expressed as follows:

$$R = P\left\{ \prod_{l=1}^{g-D-B+2} \left[I\left(\sum_{i=l}^{l+B-1} \left[I\left(\sum_{j=i}^{i+D-1} b_j < w \right) \right] < A \right) \right] = 1 \right\}. \tag{10.4}$$

With the help of equation (10.4), we can evaluate the signature of the system having i.i.d. components as $s_A = p_D(T_s = T_{A:g})$, where T is the system lifetime and s_A is the probability of the system failure.

Boland [32] obtained the structure function R of the system having i.i.d. components as follows:

$$s_A = \frac{1}{\binom{g}{g-A+1}} \sum_{l \subseteq [g]} \varphi(R) - \frac{1}{\binom{g}{g-A}} \sum \varphi(R). \tag{10.5}$$

10.4 Proposed Algorithms

10.4.1 Algorithm for Evaluating the Reliability of a k-out-of-n SWS (See Levitin and Dai [9])

Step 1. Initialization:
$$F = 0; \quad U_{1-r}(z) = z^{0,x_0}.$$

Step 2. Compute $U_{j+1-r}(z) = U_{j-r}(z) \otimes u_j(z)$, and collect the like terms in the obtained u-function.

Step 3. If $j \geq k + r - 1$, then add $\delta(U_{j+1-r}(z))$ to F and eject all the terms t with $c_{j-r+1,t} = k$ from $U_{j+1-r}(z)$.

Step 4. Remove from $U_{j+1-r}(z)$ all the terms with $c_{j-r+1,t} < k - m + j$.

Step 5. Evaluate the reliability of a k-out-of-n SWS as $R = 1 - F$.

10.4.2 Algorithm for Calculating the Signature of a k-out-of-n SWS with its Reliability Function

Step 1. Calculate the system signature of the structure function (Boland [32]).

$$B_l = \frac{1}{\binom{m}{m-n+1}} \sum_{\substack{H \subseteq [m] \\ |H|=m-n+1}} \phi(H) - \frac{1}{\binom{m}{m-n}} \sum_{\substack{H \subseteq [m] \\ |H|=m-n}} \phi(H). \quad (10.6)$$

Evaluate reliability polynomial of a k-out-of-n SWS by

$$H(P) = \sum_{\varepsilon=1}^{m} C_e \binom{m}{e} p^\varepsilon q^{n-\varepsilon}$$

where $C_j = \sum_{j=m-e+1}^{m} B_j, e = 1, 2, ..., m.$

Step 2. Compute the tail signature of a k-out-of-n SWS, i.e., $(m+1)$-tuple $B = (B_0, ..., B_m)$ using

$$B_l = \sum_{j=a+1}^{m} b_j = \frac{1}{\binom{m}{m-a}} \sum_{|H|=m-a} \phi(H). \quad (10.7)$$

Step 3. Evaluate the reliability function in the form of a polynomial by using Taylor expansion about $x = 1$ by

$$P(x) = y^m h\left(\frac{1}{y}\right). \tag{10.8}$$

Step 4. Assess the tail signature of the *k*-out-of-*n* SWS reliability function with the help of equation (10.6) (see Marichal and Mathonet [27]):

$$B_a = \frac{(m-a)!}{m!} D^a P(1), \ a = 0, \ldots, m. \tag{10.9}$$

Step 5. Obtain the signature of the *k*-out-of-*n* SWS using equation (10.8) as follows:

$$b = B_{a-1} - B_a, \ a = 1, \ldots, m. \tag{10.10}$$

10.4.3 Algorithm to Assess the Expected Lifetime of a *k*-out-of-*n* SWS with Minimum Signature

Step 1. Determine the expected lifetime of an i.i.d. component *k*-out-of-*n* SWS which are exponentially distributed with mean $\mu = 1$.

Step 2. Calculate the minimum signature of a *k*-out-of-*n* SWS with the expected lifetime of the reliability function by using

$$\bar{h}_T(t) = \sum_{j=1}^n C_j h_{1:j}(t) \tag{10.11}$$

where $h_{1:j}(t) = P_r(Z_{1:j} > t)$ and $h_{j:j}(t) = P_r(Z_{j:j} > t)$ for $j = 1, 2, \ldots, n$.

Step 3. Obtain $E(T)$ of a *k*-out-of-*n* SWS of i.i.d. components by (see Navarro and Rubio [33]).

$$E(T) = \mu \sum_{j=1}^n \frac{C_j}{j} \tag{10.12}$$

where $C_j (j = 1, \ldots, n)$ is a vector coefficient of minimal signature.

10.4.4 Algorithm for Evaluating the Expected Value of the Component *X* and Expected Cost Rate of a *k*-out-of-*n* SWS When Working Elements Are Failed

Step 1. Calculate the number of failed elements at the time of system failure with signature (Eryilmaz [34]):

$$E(X) = \sum_{j=1}^n j \cdot b_j, \ j = 1, 2, \ldots, n. \tag{10.13}$$

Step 2. Compute the $E(X)$ and $E(X)/E(T)$ of a k-out-of-n SWS with minimum signature.

10.5 Illustration

Consider a 2-out-of-3 SWS for $m = 4$, $r = 2$, $w = 3$ and each window has two states: completely working and total failure along with some performance rate 1, 2, 2, and 2, respectively. A diagram of the proposed system is shown in Figure 10.1.

The probability of the inner parallel component can be expressed as follows:

$$P_e = 1 - \prod_{m=1}^{n}(1 - R_{em})$$

where $e = 1, 2, 3, 4$, $m = 1, 2$.

The probability P_e, $e = 1, 2, 3, 4$, of the parallel system is obtained as follows:

$$P_e = R_{e1} + R_{e2} - R_{e1}R_{e2}. \tag{10.14}$$

Now, the u-function of the k-out-of-n SWS is given by

$$U_e(z) = P_e z^e + (1 - P_e)z^0$$

where $e = 1, 2, 3, 4$, P_e is the probability function and z^e is the performance rate and z^0 non performance rate.

Thus, the u-function of the k-out-of-n SWS components $u_i(z)$ is given by

$$u_e(z) = P_e z^a + (1 - P_e)z^0$$

where $a = 1, 2, 2, 2$.

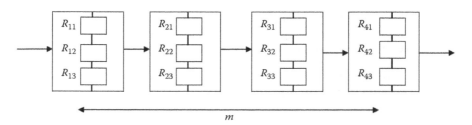

FIGURE 10.1
Diagram of k-out-of-n SWS with $k = 2$, $n = 3$, $m = 4$, and $r = 2$.

Signature Reliability of k-out-of-n

In the initial step of the algorithm, the value of 0 is assigned to F. The initial u-function takes the form

$$U_{-1}(z) = z^{0,(0,0)}.$$

Using the algorithm 10.4.1, we get the u-function of the k-out-of-n SWS as follows:

For $j = 1$

$$U_0(z) = (U_{-1}(z) \otimes u_1(z))$$
$$= z^{0,0} \otimes P_1 z^1 + (1-P_1)z^0$$
$$= P_1 z^{0,(0,1)} + (1-P_1) z^{0,(0,0)}$$

For $j = 2$

$$U_1(z) = U_0(z) \otimes u_2(z)$$
$$= P_1 z^{0,(0,1)} + (1-P_1) z^{0,(0,0)} \otimes P_2 z^2 + (1-P_2) z^0$$

$$U_1(z) = P_1 P_2 z^{0,(1,2)} + P_1(1-P_2) z^{1,(1,0)} + P_2(1-P_1) z^{1,(0,2)} + (1-P_1)(1-P_2) z^{1,(0,0)}$$

For $j = 3$

$$U_2(z) = U_1(z) \otimes u_3(z)$$
$$= P_1 P_2 z^{0,(1,2)} + P_1(1-P_2) z^{1,(1,0)} + P_2(1-P_1) z^{1,(0,2)}$$
$$+ (1-P_1)(1-P_2) z^{1,(0,0)} \otimes P_3 z^2 + (1-P_3) z^0$$

$$U_3(z) = P_1 P_2 P_3 z^{0,(2,2)} + P_1(1-P_2) P_3 z^{2,(0,2)} + (1-P_1) P_2 P_3 z^{1,(2,2)}$$
$$+ (1-P_1)(1-P_2) P_3 z^{2,(0,2)} + P_1 P_2 (1-P_3) z^{1,(2,0)}$$
$$+ P_1(1-P_2)(1-P_3) z^{2,(0,0)} + (1-P_1) P_2 (1-P_3) z^{2,(2,0)}$$
$$+ (1-P_1)(1-P_2)(1-P_3) z^{2,(0,0)}.$$

The like terms in $U_2(z)$ are collected. The value of unreliability F can be expressed as follows:

$$F = (1-P_2)P_3 + (1-P_1)P_2(1-P_3) + (1-P_2)(1-P_3). \tag{10.15}$$

After removing the terms in which the counter equals $k = 2$, $U_2(z)$ can be written as follows:

$$U_2(z) = P_1 P_2 P_3 z^{0,(2,2)} + (1-P_1) P_2 P_3 z^{1,(2,2)} + P_1 P_2 (1-P_3) z^{1,(2,0)}.$$

After removing the terms that satisfy the condition (10.3) for $k = 2$, $m = 4$, and $j = 3$,
$U_2(z)$ can be obtained as follows:

$$U_2(z) = (1 - P_1) P_2 P_3 z^{1,(2,2)} + P_1 P_2 (1 - P_3) z^{1,(2,0)}.$$

For $j = 4$

$$U_3(z) = U_2(z) \otimes u_4(z)$$

$$= (1 - P_1) P_2 P_3 z^{1,(2,2)} + P_1 P_2 (1 - P_3) z^{1,(2,0)} \otimes P_4 z^2 + (1 - P_4) z^0$$

$$U_3(z) = (1 - P_1) P_2 P_3 P_4 z^{1,(2,2)} + P_1 P_2 (1 - P_3) P_4 z^{2,(0,2)}$$

$$+ (1 - P_1) P_2 P_3 (1 - P_4) z^{2,(2,0)} + P_1 P_2 (1 - P_3)(1 - P_4) z^{2,(0,0)}.$$

Again like terms in $U_3(z)$ are collected. The value of unreliability F can be expressed as follows:

$$F = P_1 P_2 (1 - P_3) P_4 + (1 - P_1) P_2 P_3 (1 - P_4) + P_1 P_2 (1 - P_3)(1 - P_4). \quad (10.16)$$

Now adding the equations (10.15) and (10.16), we finally get the unreliability:

$$F = 1 - P_1 P_2 P_3 - P_2 P_3 P_4 + P_1 P_2 P_3 P_4$$

Reliability of the k-out-of-n SWS:

$$R = 1 - F = P_1 P_2 P_3 + P_2 P_3 P_4 - P_1 P_2 P_3 P_4. \quad (10.17)$$

Hence, substituting the values of P_e ($e = 1, 2, 3, 4, 5$) in equation (10.17) from equation (10.14), we obtain the reliability function R of the k-out-of-n SWS as follows:

$$R = (R_{11} + R_{12} - R_{11} R_{12})(R_{21} + R_{22} - R_{21} R_{22})(R_{31} + R_{32} - R_{31} R_{32})$$

$$+ (R_{21} + R_{22} - R_{21} R_{22})(R_{31} + R_{32} - R_{31} R_{32})(R_{41} + R_{42} - R_{41} R_{42})$$

$$- (R_{11} + R_{12} - R_{11} R_{12})(R_{21} + R_{22} - R_{21} R_{22})(R_{31} + R_{32} - R_{31} R_{32})$$

$$\times (R_{41} + R_{42} - R_{41} R_{42}). \quad (10.18)$$

When elements are identically distributed ($R_{em} \equiv R$), reliability function $R(R_1, ..., R_8)$ of the components of the k-out-of-n SWS and structure function h of the proposed system are given by

$$R(R_1, ..., R_8) = 16R^3 - 40R^4 + 44R^5 - 26R^6 + 8R^7 - R^8$$

Signature Reliability of k-out-of-n 243

and
$$H(P_1,\ldots,P_8) = 16P^3 - 40P^4 + 44P^5 - 26P^6 + 8P^7 - P^8.$$

10.5.1 Signature of the *k*-out-of-*n* SWS

By using Owen's method on the components of the *k*-out-of-*n* SWS, we get reliability function in the form of $H(y)$ as follows:

$$H(y) = 16y^3 - 40y^4 + 44y^5 - 26y^6 + 8y^7 - y^8. \tag{10.19}$$

Now using equations (10.8) and (10.19), structure function can be expressed as follows:

$$P(y) = y^8 H\left(\frac{1}{y}\right) = -1 + 8y - 26y^2 + 44y^3 - 40y^4 + 16y^5.$$

With the help of step 4 of algorithm 10.4.2, we get the tail signature *B* for individual element of the *k*-out-of-*n* SWS as follows:

$$B_0 = 1, B_1 = 1, B_2 = \frac{13}{14}, B_3 = \frac{11}{14}, B_4 = \frac{4}{7},$$

$$B_5 = \frac{2}{7}, B_6 = 0, B_7 = 0, B_8 = 0.$$

Hence, the tail signature of the *k*-out-of-*n* SWS is given by

$$B = \left(1, 1, \frac{13}{14}, \frac{11}{14}, \frac{4}{7}, \frac{2}{7}, 0, 0, 0\right).$$

Again using step 5 of algorithm 10.4.2, we obtain the signature of the *k*-out-of-*n* SWS as follows:

$$b = \left(0, \frac{1}{14}, \frac{1}{7}, \frac{3}{14}, \frac{2}{7}, \frac{2}{7}, 0, 0\right).$$

10.5.2 MTTF of the *k*-out-of-*n* SWS

To compute MTTF using equation (10.19), we get a minimal signature *M* as follows:
 $M = (0, 0, 16, -40, 44, -26, 8, -1)$ of the *k*-out-of-*n* SWS elements.
 Using the steps 2 and 3 of algorithm 10.4.3, we get MTTF

$$E(t) = 0.818.$$

10.5.3 Expected Cost

Using step 1 of algorithm 10.4.4, the expected value of X of the k-out-of-n SWS can be computed as follows:

$$E(X) = \sum_{j=1}^{8} j \cdot B_j, \, j = 1, 2, \ldots, 8.$$

Hence, the expected value of X is given by

$$E(X) = 4.57.$$

Now, using step 2 of algorithm 10.4.4, we can compute the expected cost of the k-out-of-n SWS as follows:

$$\text{Expected cost} = \frac{E(X)}{E(T)} = 5.5867.$$

10.6 Conclusion

In this study, we considered a k-out-n SWS consisting of n linearly ordered multistate components with m parallel components. We evaluated signature, tail signature, expected lifetime, and expected cost. Signatures increased with increasing parallel component, expected lifetime was 0.818, and expected cost was 5.5867.

Nomenclature

n = number of multistate element (MSE) in the system
m = consecutive i.i.d. components
r = number of consecutive window
w = total allocation weight
k = maximal allocation consecutive groups.
$U_a(z)$ = u-function of the r consecutive MSE a
$u_i(z)$ = u-function of the system
$\otimes$ = composition operator
$g_{i,b}$ = the performance state of MSE in state b
$y_{a,l}$ = random vector in lth state of the r consecutive groups
$C_{a,l}$ = integer counter of consecutive failed groups

$P_{i,b}$ = the probability of MSE i is in state b

$q_{i,b}$ = probability in bth state of the r consecutive groups

E_a = probability of m consecutive groups

$\sigma(y)$ = sum of element vector y

$\phi(y, g)$ = shifting operator

$R/F/S/s/H$ = reliability/ unreliability /tail signature/signature/reliability function of the system

$E(T)/E(X)$ = expected lifetime/expected value of X of the system components

C_e = minimum signature for components e

References

1. Chiang, D. T. & Niu, S. C. (1981). Reliability of consecutive k-out-of-n: F system. *IEEE Transactions on Reliability*, 30(1), 87–89.
2. Levitin, G. (2003). Linear multi-state sliding-window systems. *IEEE Transactions on Reliability*, 52(2), 263–269.
3. Koucký, M. (2003). Exact reliability formula and bounds for general k-out-of-n systems. *Reliability Engineering & System Safety*, 82(2), 229–231.
4. Levitin, G. (2005). Reliability of linear multistate multiple sliding window systems. *Naval Research Logistics (NRL)*, 52(3), 212–223.
5. Habib, A., Al-Seedy, R. O., & Radwan, T. (2007). Reliability evaluation of multistate consecutive k-out-of-r-from-n: G system. *Applied Mathematical Modelling*, 31(11), 2412–2423.
6. Ram, M. & Singh, S. B. (2009). Analysis of reliability characteristics of a complex engineering system under copula. *Journal of Reliability and Statistical Studies*, 2(1), 91–102.
7. Levitin, G. & Ben-Haim, H. (2011). Consecutive sliding window systems. *Reliability Engineering & System Safety*, 96(10), 1367–1374.
8. Levitin, G. & Dai, Y. (2011). Linear m-consecutive k-out-of-r-From-n: F systems. *IEEE Transactions on Reliability*, 60(3), 640–646.
9. Levitin, G. & Dai, Y. (2012). k-out-of-n sliding window systems. *IEEE Transactions on Systems, Man, and Cybernetics-Part A: Systems and Humans*, 42(3), 707–714.
10. Xiang, Y. & Levitin, G. (2012). Combined m-consecutive and k-out-of-n sliding window systems. *European Journal of Operational Research*, 219(1), 105–113.
11. Ram, M. & Singh, S. B. (2012). Cost benefit analysis of a system under head-of-line repair approach using Gumbel-Hougaard family copula. *Journal of Reliability and Statistical Studies*, 5(2), 105–118.
12. Ram, M. (2013). On system reliability approaches: A brief survey. *International Journal of System Assurance Engineering and Management*, 4(2), 101–117.
13. Pham, H. (2014, October). Computing the reliability of complex systems. In *2014 3rd International Conference on Reliability, Infocom Technologies and Optimization (ICRITO) (Trends and Future Directions)*, IEEE, Noida, p. 1.

14. Negi, S. & Singh, S. B. (2015). Reliability analysis of non-repairable complex system with weighted subsystems connected in series. *Applied Mathematics and Computation, 262,* 79–89.
15. Ram, M. & Davim, J. P. (Eds). (2017). *Advances in Reliability and System Engineering.* Springer International Publishing, Cham, Switzerland.
16. Shapley, L.S. (1953). A value for n-person games. In H. W. Kuhn and A. W. Tucker (Eds.), In *Contributions to the Theory of Games, Vol. 2. Annals of Mathematics Studies,* Vol. 28. Princeton University Press, Princeton, NJ, pp. 307–317.
17. Owen, G. (1972). Multilinear extensions and the Banzhaf value. *Naval Research Logistics Quarterly, 22*(4), 741–750.
18. Owen, G. (1988). Multilinear extensions of games. In A.E. Roth, In The Shapley Value. Essays in Honor of Lloyd S. Shapley, Cambridge University Press, Cambridge, pp. 139–151.
19. Samaniego, F. J. (1985). On closure of the IFR class under formation of coherent systems. *IEEE Transactions on Reliability, 34*(1), 69–72.
20. Kochar, S., Mukerjee, H., & Samaniego, F. J. (1999). The signature of a coherent system and its application to comparisons among systems. *Naval Research Logistics, 46*(5), 507–523.
21. Boland P.J., Samaniego F.J. (2004) The Signature of a Coherent System and Its Applications in Reliability. In: Soyer R., Mazzuchi T.A., Singpurwalla N.D. (eds) Mathematical Reliability: An Expository Perspective. International Series in Operations Research & Management Science, vol 67. Springer, Boston, MA.
22. Samaniego, F. J. (2007). *System Signatures and Their Applications in Engineering Reliability,* Vol. 110. Springer Science & Business Media, USA.
23. Navarro, J. & Rychlik, T. (2007). Reliability and expectation bounds for coherent systems with exchangeable components. *Journal of Multivariate Analysis, 98*(1), 102–113.
24. Navarro, J., Samaniego, F. J., Balakrishnan, N., & Bhattacharya, D. (2008). On the application and extension of system signatures in engineering reliability. *Naval Research Logistics (NRL), 55*(4), 313–327.
25. Samaniego, F. J., Balakrishnan, N., & Navarro, J. (2009). Dynamic signatures and their use in comparing the reliability of new and used systems. *Naval Research Logistics (NRL), 56*(6), 577–591.
26. Da, G., Zheng, B., & Hu, T. (2012). On computing signatures of coherent systems. *Journal of Multivariate Analysis, 103*(1), 142–150.
27. Marichal, J. L. & Mathonet, P. (2013). Computing system signatures through reliability functions. *Statistics & Probability Letters, 83*(3), 710–717.
28. Coolen, F. (2013). System reliability using the survival signature. *Survival, 1,* 31.
29. Kumar, A. & Singh, S. B. (2017). Computations of signature reliability of coherent system. *International Journal of Quality & Reliability Management, 34*(6), 785–797.
30. Kumar, A. & Singh, S. B. (2017). Signature reliability of sliding window coherent system. In Mangey Ram and J.P. Davim (Eds.), In *Mathematics Applied to Engineering,* Elsevier International Publisher, London, pp. 83–95.
31. Kumar, A. & Singh, S. B. (2017). Signature reliability of linear multi-state sliding window system. *International Journal of Quality & Reliability Management,* (Accepted).
32. Boland, P. J. (2001). Signatures of indirect majority systems. *Journal of Applied Probability, 38*(2), 597–603.

33. Navarro, J. & Rubio, R. (2009). Computations of signatures of coherent systems with five components. *Communications in Statistics-Simulation and Computation*, 39(1), 68–84.
34. Eryilmaz, S. (2012). The number of failed components in a coherent system with exchangeable components. *IEEE Transactions on Reliability*, 61(1), 203–207.

11

Modeling Reliability of Component-Based Software Systems

Preeti Malik, Lata Nautiyal, and Mangey Ram
Graphic Era University

CONTENTS

11.1 Introduction 249
11.2 Software Failure Mechanisms 251
11.3 Software versus Hardware Reliability 252
11.4 Component-Based Software Development 254
    11.4.1 Characteristics of Component by Lau et al. [16] 255
    11.4.2 Reliability of a Component 256
    11.4.3 Reliability of Component-Based Software 256
    11.4.4 State- and Path-Based Models 257
        11.4.4.1 State-Based Models 258
        11.4.4.2 Path-Based Models 260
    11.4.5 Use of CBD in Traditional Manufacturing 261
11.5 Conclusion 262
References 263

11.1 Introduction

"Reliability is the precondition for trust."

Wolfgang Schauble

The transition of current infrastructure, the revaluation of a collaborated or incorporated strategy, or the change of ownership and acquisition may be some of the factors for the continual requirement of homogenization and alteration of particular software. This is the driving force of all evolution of business applications. Rightfully termed "legacy applications," these improvised applications are expected and conjectured to revolutionize software design and architecture as never before. In computing, a legacy system is an old methodology which links to or relates to an outdated, predecessor computer. Reliability assessment of upgraded legacy systems is an important problem in IT software infrastructure. Some parts of the code used in the original design of such systems are currently being discontinued. Maintaining a legacy system, therefore, demands upgradation of the software constituents. The most necessary requirement after upgrading an application is the assurance of software reliability. Tests of reliability are supposed to be conducted on this software, for which the best-known approach is the Bayesian approach.

Previous studies have made use of this Bayesian theory under predefined assumptions. This opens the door for future research after thorough analysis of the predictions of the Bayesian approach. Software engineering is a coherent, assessable, measurable, and methodical approach for the evolution, perpetuation, performance, and working of a software application. It is now an established vocation and is committed toward developing software that is economical, easier to maintain, quicker, and of higher quality. Since the field is still relatively young compared to its relative fields of engineering, there is still much work and debate around what software engineering (SE) actually is, and if it deserves the term "engineering." It has grown organically out of the limitations of viewing software as just programming. Software development is a term sometimes preferred by practitioners in the industry who view SE as too heavy handed and constrictive to the integrated process of creating software. The software development life cycle (SDLC) is a general term used in software engineering, which constitutes the five software developing actions such as *planning, creating, testing, deploying,* and *maintaining* an information system.

The development of a computer programming language can be understood by measuring the complexity of computer programs with respect to the size of the programs. Another way of looking at the evolution of programming languages is getting the computer to accomplish increasingly complex tasks. Lack of understanding of a program's overall structure and functionality will result in the failure of detecting errors in the program. This can be avoided by using better languages that conversely reduce the number of errors by enabling a better understanding. At even greater levels of abstraction, this is what the attempted software design will provide. The involvement of subroutines, statements, files, classes, templates, and other such components permits certain sections of the program to be abstracted. Layers, hierarchies, and modules help achieve comprehensibility of the code. Also, advances in languages provide the engineers with more control over shape and use of data elements as abstract types. These data types are very accurately and finely specified.

Software plays a key role in the modern world. The value of software is derived from its ability to increase productivity and efficiency, its resiliency to attack, and its ability to perform at the required levels during the time of crisis and normal operations. The development of successful software applications requires an engineering approach which is categorized by the application of scientific theories, methodologies, models, and standards which make it possible to manage, plan, analyze, model, design, implement, maintain, measure, and evolve a software system. The famous definition of software reliability is: "It is the probability of the failure less performance of the software over a given period of time" [1].

Though software reliability is also expressed as a hypothetical function and shown in the concept of time, it must be noted that it is completely different from conventional hardware reliability. Both the reliabilities are totally different, as one is in tangible form and another is in intangible form. Hardware is made up of various electronic and mechanical parts, which may become "old" after a certain period of time and wear out with time. When it is the matter of software reliability, it will not rust or wear out during its complete life span. Software will remain as it is till it is not intentionally changed. Software reliability is one of the essential features of software quality, along with a large number of important elements such as performance, serviceability, functionality, usability, and documentation. It is very difficult to achieve software reliability as the software systems are highly complex; therefore, it is the most important quality of the software system. Software systems, or any systems which are known to have high degrees of complexities and convolutions, have to strive harder to attain an established level of definition and reliability. Hence, system developers are inclined toward pushing intricacies and complexities into the software layer. The process thus gets easier due to the rapid growth of the system and the upgradation and enhancement of the software [2–4].

The complexity of software is directly dependent on and related to factors like software quality in terms of capability, functionality, etc., and is inversely related to software reliability. Nowadays, software reliability engineering has become very important [5]. There are various models which are used for calculating software reliability. A major issue in reliability engineering is assessing the reliability of the software application. However, it is not as easy as it seems. The biggest obstacle is with respect to design faults and this is handled differently than the conventional hardware theory. A fault is nothing but an error in the code made by the programmer or designer with regard to the specification of the software. If an input value activates a fault, it brings an incorrect output and once this happens, it will result in software failure. Software failure is detected using stochastic models of stochastic processers which govern software failures [6,7].

11.2 Software Failure Mechanisms

Ambiguities, errors, oversights, misinterpretations, carelessness, and incompleteness in code lead to incorrect or unexpected use of the software, or other unforeseen problems. Though almost always stated otherwise, software reliability and hardware reliability both have entirely different processes of failure. Faults associated with hardware will always be physical or tangible, whereas all the faults associated with software will be design

TABLE 11.1

A partial list of the distinct characteristics of software compared to hardware.

1	Reliability forecast
2	Failure reason
3	Environmental influences
4	Wear-out
5	Time reliance and life cycle
6	Idleness

related [8–10]. All the faults related to software are very difficult to imagine, detect, as well as correct. Design faults may also exist in hardware faults, but the dominance is toward the physical faults. Table 11.1 shows the partial list of the distinct characteristics of software compared to hardware.

11.3 Software versus Hardware Reliability

The fundamental differences between hardware and software failures are defined by Reliability Analysis Centre [11]. In hardware, the failures are usually caused by physical processes related to stresses forced by the environment. Generally, the failures are due to component degradation, failing, or being subjected to environmental stresses. In software, there is nothing to wear out. Software system failures never happen if the software is not used. This is not true for hardware systems where material worsening causes failures even though the system is not being used. Software reliability models are usually analytical models derived from the assumptions of the system, and the elucidation of those assumptions with model parameters. The dispensations from failure data derive techniques of reliable hardware. Extensive scrutiny, analysis, and experience in this particular field help achieve this. Then, a new version of a software system is obtained if the faults in a software system are being repaired. This is not true for hardware repairs that typically restore the original system. A massive keynote here is that hardware reliability changes dynamically, throughout its lifetime, whereas software reliability is continuously upgraded and refined until its delivery. This can be graphically represented in Figures 11.1 and 11.2.

The degree, up to which the results of measurement, calculation, specification, or statistics can be depended on the accuracy, is known as reliability. The four elements that reliability establishes are:

i. Intended function
ii. Time

Component-Based Software Systems

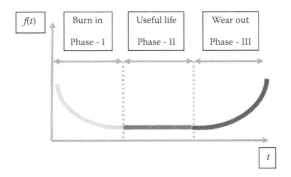

FIGURE 11.1
Hardware failure rate.

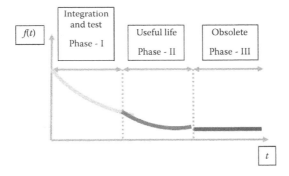

FIGURE 11.2
Software failure rate.

iii. Conditions of operation

iv. Probability

Let T be the time until occurrence of failure of unit. Hence, the probability that the unit will not fail in a particular domain until time t is:

$$R(t) = P(T > t) \tag{11.1}$$

where $R(t)$ is the reliability function.

We can hence state that reliability is always a function of time. The conditions it depends on may or may not vary with time. The numerical value of reliability is always between 0 and 1. That is, $R(t)$ is a nonincreasing function and its limits are 0 to 1.

Noteworthy definitions from various editions of IEEE:

- IEEE 1998 defines software reliability as the ability of a system or a component to perform its required functions under stated conditions for a specified period of time.

- IEEE 1988 defines software reliability as the process of optimizing the reliability of software through a program that emphasizes software error prevention, detection of fault and its removal, and usage of measurements in order to maximize reliability in aspects of resources, performance, and schedule.

This brings to us the three main factors of software reliability:

1. Prevention of error
2. Detection and removal of faults
3. Measurements to maximize reliability

11.4 Component-Based Software Development

The component-based software development (CBSD) approach aims to develop software systems using existing components assembled with well-structured software architecture. Its purpose is to reduce time and capital invested in the development of software. CBSD claims to reduce time and energy on maintenance, stating that certain sections of a software application may be written only once and henceforth reused time and again. CBSD personifies the "buy, don't build" philosophy and aims at discovering reuse of software by changing both architecture and process. CBSD incorporates two collateral activities of engineering:

  i. Domain engineering
  ii. Component-based development

CBSD is different from orthodox approaches, which do not use the idea of recycling the code. Figure 11.3 depicts commercial-off-the-shelf components, developed by various engineers, using different languages, assembled into a targeted software system.

Kaur [11] determines the following four traits of CBSD:

  i. **Independent development of software:** Disassociate developers and users of components through abstract, conceptual interface specifications of component behavior.
  ii. **Reusability:** Certain preexisting sections of components can be reused.
  iii. **Quality of software:** Quality assurance approach of software must level from modular to scalable.
  iv. **Maintenance:** System software must be comprehensible and easy to evolve.

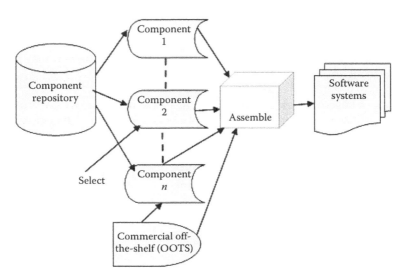

FIGURE 11.3
Component-based software development.

We need to establish a difference between a component and a software component.

D. F. D'Souza [12] defines a component as a reasoned package of software that may be developed and henceforth delivered independently as a unit, among other such components, to build a larger system.

C. Szyperski [13] defines a software component as a unit of configuration with conventionally specified interfaces, which can be deployed independently and is subject to configuration by third-party organizations.

G. T. Heineman [14] establishes a definition of a component which states that a component is an element of software which conforms to a model and can be separately situated and formulated without being modified according to a composition standard.

Meyer [15] states that a component is a unit of software that satisfies the following specifications:

i. Components can be used by clients
ii. Components possess official usage description
iii. Components do not have a fixed set of clients

11.4.1 Characteristics of Component by Lau et al. [16]

- A component is a data capsule which hides fundamental information.
- A component can be administered in any language, and is not just object or module oriented.

- The module of the components can be described either textually or visually.
- The component framework architecture forms its plug and play software technology.

Component-based development (CBD) has one major goal—to build and maintain software using existing components. Component-based software engineering (CBSE) states four attributes of a truly reusable component:

a. Interfaces to be contractually specified
b. Autonomic/independent deployment
c. Explicit context dependence
d. Third-party configuration

11.4.2 Reliability of a Component

We know that a system is made up of different parts and components. If one single component fails, the entire system is vulnerable to failure. Each component subscribes to system reliability. Component failure models are obtained in two ways:

i. By the basic failure rates and working stresses of the system
ii. Using component failure data obtained either from life tests or from failure reports of customers

It is impossible to establish the failure rates of components for all working conditions. It is possible to predict the reliability under specific stress conditions from the available data.

11.4.3 Reliability of Component-Based Software

CBSE has now become a more widespread approach for the development of software systems. The advantages of the CBSD approach are low cost, high quality, minimal efforts, and reduced development time. They are mainly because of predefined software components. Many new models and techniques are proposed for effectively improving the reliability estimation of component-based software systems. However, ensuring a reliable component-based software system is difficult even with the use of commercial pretested and trusted software components. Thus, the reliability evaluation of component-based software systems is essential in the present scenario of software usage either before or after implementation.

CBD is a unique way to form, construct, implement, and calculate software applications. Various sources are considered in collecting and depicting these software applications; different languages are used to develop the component which works in different steps [17,18]. The main concept behind CBD is "reuse." Software productivity can be increased with software product reuse because reused software components need not to be developed from scratch. Component-based software can save time to market those results in larger market share.

With software reuse, prototypes can be developed very quickly and at very low cost as instead of developing specifications, designs, and code from scratch, existing components can be reused. The current trend in software engineering is toward CBD. CBSE is a promising alternative to enhance software productivity. It can improve quality and reliability of the software. It may help to identify design errors at early stage by plugging different candidate components. Still software reuse is not in practice to its potential due to a number of reasons.

Software organizations face problem in practicing software reuse because of the following reasons:

a. Object-oriented analysis, computer-aided software engineering tools, formal methods, agile methods, etc., are proposed but no clear-cut methodology for software reuse is defined.
b. There is an absence of models and theories that can help to comment or estimate about the software without developing it.
c. Rapid evolution of technologies does not allow proper assessment.
d. Collecting context-independent knowledge or representing context is very difficult.
e. Software development for reuse especially for large complex systems is very cumbersome.
f. There is lack of empirical studies and technologies that can help to estimate what a component can perform better and what it cannot, without using it.

The main problem in software reuse is the effective design and development of a software life cycle model, performing a software testing process, and also the absence of an effective component-based certification process for reusable components. Most of the existing CBSE models, testing techniques, and certification processes are either domain specific or very difficult to use. What derives from the above discussion is that further research needs to be done to come up with more effective reusable software component CBSE models, testing techniques, and reliability processes for CBSE.

11.4.4 State- and Path-Based Models

Black box and white box are two approaches for modeling reliability of component-based software systems. The white box approach considers internal behavior of the system. Basically, these approaches can further be classified into two categories: state- [19] and path-based modeling [20]. The second-category approach generally assumes that components do not depend on each other for execution. These approaches are also known as pessimistic approaches.

11.4.4.1 State-Based Models

State-based models estimate software reliability analytically. They assume that the transfer of control between modules has a Markov property, that is, model software architecture with a discrete-time Markov chain (DTMC), a continuous-time Markov chain (CTMC), or a semi-Markov process (SMP). Below we will discuss some of the leading literature on state-based models.

    a. Littewood model: This was one of the earliest approaches to estimate software reliability. It considers software reliability in terms of operational reliability. First, a reliability system with system architecture based upon irreducible CTMC was made [21. Another approach [22] was developed that consists of a modular program in which the transfer of control between modules follows an SMP. This model describes structure via dynamic behavior using Markov assumption. It analyzes both the component and interface failures. The Littlewood model is a nearly universal architecture-based model. Irreducible SMP is used to model the architecture of component-based software (CBS). This model presumes that CBS is an integration of a number of modules and a control is transferred from one module to another with probability p_{ij}, where [23]

$$p_{ij} = P_r\{\text{Program transit from } i\text{th model to } j\text{th module}\}.$$

A general distribution function F_{ij} is employed to describe the time used up in a module and the mean is m_{ij} [23]. Two types of failure behavior are considered in this model:

- λ_i is the failure during execution of the component, and
- v_{ij} is the failure during transfer of control between two components.

The combined failure rate is

$$\lambda_s = \sum_i a_i * \lambda_i + \sum_{i,j} b_{ij} * V_{ij} \quad (11.1)$$

where $a_i = \dfrac{\Pi_i \sum_j p_{ij} * m_{ij}}{\sum_i \Pi_i * \sum_j p_{ij} * m_{ij}}$ and $b_{ij} = \dfrac{\Pi_i * p_{ij}}{\sum_i \Pi_i * \sum_j p_{ij} * m_{ij}}$.

b. Laprie model [24]: This model is a special case of the Littlewood model in that it considers only component failure. This model says that the software system follows CTMC. The Laprie model takes the constant failure rate to be λ_i and the time spent in each component is μ_i. λ_i is much smaller than μ_i. The system failure rate can be defined as, according to Laprie:

$$\lambda_s = \sum_{i=1}^{n} \Pi_i * \lambda_i \qquad (11.2)$$

where n is the component count, π_i is the ratio of time used up in ith component, and λ_i is the failure rate of component i [23].

c. Cheung model [19]: This model makes use of DTMC. The reliability of the service provided by the system is measured by a user-oriented model. Transition probability is also taken into account by Cheung as a user profile. The author assumes that a particular component may execute infinite time till the termination of the execution of the system.

Let L be the set of processes P_i that can be generated by the program corresponding to different input values. Let r_i be a random variable such that

$$r_i = \begin{cases} 1 \text{ if the process } P_i \text{ generates the correct program output} \\ 0 \text{ otherwise} \end{cases}$$

Let q_i be the probability that P_i will be generated in a user environment. The values of q_i, therefore, define the user profile. The reliability R of the program can be computed from

$$R_i = \sum_{\forall P_i \in L} q_i r_i \qquad (11.3)$$

d. Kubat model: Kubat [25] made some improvements in the Cheung model. The proposed model describes the architecture of the software system as SMP by considering the execution time of the components. The DTMC process is followed by transition between components where $q = [q_i]$ is the initial probability vector and $P = [p_{ij}]$ is the transition probability matrix. The author assumes that no component fails during the execution of the software system.

e. Gokhale model [26]: The architecture of the software system is described by DTMC. This model utilizes a hierarchical solution. According to Gokhale, the reliability of the component can be described as follows:

$$R_i = e^{-\int_0^{V_i t_i} \lambda_i(t) dt} \qquad (11.4)$$

where $\lambda_i(t)$ is the time-dependent failure intensity and $V_i t_i$ is the cumulative expected time spent in the component per execution of the application.

f. Ledoux model [27]: Ledoux attempted to overcome some constraints of the well-known Littlewood model [22] for modular systems. It allows the evaluation of various dependability metrics, in particular, of availability measures. The Ledoux model is a general model that is specifically developed for software systems.

11.4.4.2 Path-Based Models

Path-based models take into account only a fixed number of component execution traces. It usually corresponds to system test cases. Following are some of the path-based models:

a. Shooman model [28]: The Shooman model is among earliest path-based models. Shooman considers frequencies of run of various paths. This model assumes that the number of paths taken by execution is fixed. The frequency of occurrence of each path and its failure probability are assumed to be known.

b. Krishnamurthy and Mathur model [20]: This model is a very simple model to measure the reliability of the system which assumes that overall reliability of the system is the product of the components visited in a particular path. It assumes that the components are independent. The estimate of component-based reliability of a program P with respect to a test set T is given by

$$R_c = \frac{\sum_{\forall t \in T} R_c^t}{|T|} \quad (11.5)$$

where the reliability of the path is P traversed; when P is executed on test-case $t \in T$ is given by

$$R_c^t = \prod_{\forall m \in M(P,t)} R_m \quad (11.6)$$

c. Yacoub, Cukic, and Ammar model [29,30]: This model follows an algorithmic approach to measure reliability of paths. Tree traversal algorithms are used for this purpose which expand all the branches of the graph which is basically the representation of the architecture of the system. Breadth first traversal is translated as summation of reliabilities of the components which is weighted by transition probability and depth first traversal is translated as multiplication of reliabilities of the components.

d. Hamlet model [31]: This is another path-based reliability modeling approach which considers the real execution traces of component execution given the mapping from the input to the output profile. This approach also considers the matter of inaccessibility of component's usage profile. According to Hamlet, the reliability of a component can be measured as

$$R = \sum_{i=1}^{n} h_i(1 - f_i) \qquad (11.7)$$

11.4.5 Use of CBD in Traditional Manufacturing

CBD can be properly understood by an example of the manufacture of bicycles. Since the 1870s, Coventry was the place where bicycles were manufactured, using the collection parts from different industrial towns and cities. After many years, the British motor car industry came into force from this manufacturing base. The implementation of this technology of software development is very popular.

In the beginning, the manufacture of these cycles was done by the local blacksmiths and mechanics of that area fulfilling a small demand. Standardization of this product created two different effects on it:

1. It flourished the importance of small workshops.
2. It also increased mass production. People living in small villages got this "homely product" manufactured at their place at a reasonable price due to less overhead costs [32]. Many big companies also participated in manufacturing of various important parts, which were supplied to both bicycle factories and local workshops.

CBD is considered to be very important in software development and management. It denotes that few requirements should be satisfied by taking components, whereas we must also consider other methods to satisfy our interest. Conventional development is an important part of CBD but it lacks few techniques and opportunities that form a complete CBD. After this process, there are different ways which came into force:

1. From conventional development
2. From extreme componentization [33,34].

A drawback can be seen in distributed computing. Few people take a distributed system as an option to a centralized system which is not distributed. It is preferred to take all systems as distributed. In our view, the centralized system is considered as a special case with only one location [35]. CBSE developing software systems offers many advantages, namely:

1. **Flexibility:** Run-time components can work independently, if designed properly. Hence, it is fair to say that component-based systems are much more adaptable and extendable than systems traditionally designed and built. Flexibility is of significance in the hardware and system software.

    The rapid and efficient migration from one operating system to another or from one database to another is the result of lower reactivity of component-based systems toward modifications and amendments. Flexibility also leads to functionality [36]. At a functional level, component-based systems are much more adaptable and extendable since they can be reused or derived from previously existing software. Ideal functionality is that which has had to be implemented just once [37,38].

2. **Reusability:** CBD enables the development of elements and constituents which entirely implement issues of technical and business aspects. Robustness, maintainability, and productivity will be supported.

3. **Maintainability:** Functionality leads to easy maintenance of software systems, which further leads to lower cost and longer life of the systems. It is also justified if one states that the difference between the terms "maintenance" and "construction" will completely disappear after a while in terms of system software.

11.5 Conclusion

In CBSD reliability assessment, component and system reliability models assess the reliabilities of individual components and the entire component-based software, respectively. Reliability is the primary attribute which defines the success and failure of the software system. In this chapter, the authors tried to accomplish some important elements related to reliability in terms of CBD. Reliability modeling can be divided into two approaches called black box and white box. This chapter discusses the second category viz. white-box approach. State based and path based are two further classifications of the white-box approach. A major drawback of the path-based approach is that they only give an estimate of the application. Paths in the architecture of the system are considered for estimating reliability of the system. Still many enhancements are needed to overcome the restrictions of the approaches discussed above.

References

1. Kumar, D., Klefsjö, B., & Kumar, U. (1992). Reliability analysis of power transmission cables of electric mine loaders using the proportional hazards model. *Reliability Engineering & System Safety*, 37(3), 217–222.
2. Goel, H. D., Grievink, J., Herder, P. M., & Weijnen, M. P. (2002). Integrating reliability optimization into chemical process synthesis. *Reliability Engineering & System Safety*, 78(3), 247–258.
3. DTO, P. (2002). *Practical Reliability Engineering*, 4th ed., John Wiley & Sons Ltd.
4. EE, C. (2000). *Reliability and Maintainability Engineering*, 1st ed., Tata McGraw-Hill Publishing Company Ltd.
5. Srinath, L. S. (1991). *Reliability Engineering*, 3rd ed., Affiliated East West Express, New Delhi.
6. Li, W., Wang, Y., & Huang, H. (2009, August). A new model for software reliability. In *Fifth International Joint Conference on INC, IMS and IDC, 2009. NCM'09*, pp. 757–760, IEEE.
7. Lyu, M. R. (2007). Software reliability engineering: A roadmap. In *29th International Conference on Software Engineering, Future of Software Engineering*, IEEE, Minneapolis, pp. 153–170.
8. Musa, J. D., & Iannino, A. (1981). Software reliability modeling: Accounting for program size variation due to integration or design changes. *ACM SIGMETRICS Performance Evaluation Review*, 10(2), 16–25.
9. Zhou, Y., & Davis, J. (2005, May). Open source software reliability model: An empirical approach. In *5-WOSSE Proceedings of the Fifth Workshop on Open Source Software Engineering*, ACM SIGSOFT *Software Engineering Notes*, Vol. 30, pp. 1–6, ACM, New York.
10. Syed-Mohamad, S. M., & McBride, T. (2008, December). A comparison of the reliability growth of open source and in-house software. In *Software Engineering Conference, 2008. APSEC'08. 15th Asia-Pacific*, pp. 229–236, IEEE.
11. Kaur, I., Sandhu, P. S., Singh, H., & Saini, V. (2009). Analytical study of component based software engineering. *World Academy of Science, Engineering and Technology*, 3(2), 304–309.
12. D'Souza, D. F. & Wills, A. C. (1998). *Objects, Components and Frameworks with UML: The Catalysis Approach*, Addison-Wesley Longman Publishing Co., Inc., Boston, MA.
13. Szyperski, C., Bosch, J., & Weck, W. (1999). Component-oriented programming. In Moreira, A. (Ed.) *Object-Oriented Technology ECOOP'99 Workshop Reader. ECOOP 1999*. Lecture Notes in Computer Science, Vol. 1743, pp. 184–192, Springer, Berlin.
14. Council, W. T., & Heineman, G. T. (2001). *Component-Based Software Engineering Putting the Pieces Together*, Addison-Wesley Longman Publishing Co., Inc., Boston, MA.
15. Meyer, B. (2003). The grand challenge of trusted components. In *Proceedings of the 25th International Conference on Software Engineering*, Portland, Oregon.
16. Lau, K. K., Ornaghi, M., & Wang, Z. (2006, November). A software component model and its preliminary formalisation. In *International Symposium on Formal Methods for Components and Objects*, pp. 1–21, Springer, Berlin, Heidelberg.

17. Rahmani, C., Siy, H., & Azadmanesh, A. (2009). An experimental analysis of open source software reliability. Department of Defense/Air Force Office of Scientific Research.
18. Karg, L. M., Grottke, M., & Beckhaus, A. (2009, December). Conformance quality and failure costs in the software Industry: An empirical analysis of open source software. In *IEEE International Conference on Industrial Engineering and Engineering Management, 2009. IEEM 2009*, pp. 1386–1390, IEEE.
19. Cheung, R. C. (1980). A user-oriented software reliability model. *IEEE Transactions on Software Engineering*, 6(2), 118–125.
20. Krishnamurthy, S., & Mathur, A. P. (1997, November). On the estimation of reliability of a software system using reliabilities of its components. In *Proceedings of the Eighth International Symposium on Software Reliability Engineering, 1997*, pp. 1–21, IEEE.
21. Littlewood, B. (1975). A reliability model for systems with Markov structure. *Applied Statistics*, 24(2), 172–177.
22. Littlewood, B. (1979). Software reliability model for modular program structure. *IEEE Transactions on Reliability*, 28(3), 241–246.
23. Goševa-Popstojanova, K., & Trivedi, K. S. (2001). Architecture-based approach to reliability assessment of software systems. *Performance Evaluation*, 45(2–3), 179–204.
24. Laprie, J. C. (1984). Dependability evaluation of software systems in operation. In *IEEE Transactions on Software Engineering*, Vol. SE-10, Issue 6, pp. 701–714, IEEE.
25. Kubat, P. (1989). Assessing reliability of modular software. *Operations Research Letters*, 8(1), 35–41.
26. Gokhale, S. S., Wong, W. E., Trivedi, K. S., & Horgan, J. R. (1998, September). An analytical approach to architecture-based software reliability prediction. In *IEEE International Computer Performance and Dependability Symposium, IPDS'98*, pp. 13–22, IEEE.
27. Ledoux, J. (1999). Availability modeling of modular software. *IEEE Transactions on Reliability*, 48(2), 159–168.
28. Shooman, M. L. (1976, October). Structural models for software reliability prediction. In *Proceedings of the 2nd International Conference on Software Engineering*, pp. 268–280, IEEE Computer Society Press.
29. Yacoub, S. M., Cukic, B., & Ammar, H. H. (1999). Scenario-based reliability analysis of component-based software. In *Proceedings 10th International Symposium on Software Reliability Engineering, 1999*, pp. 22–31, IEEE.
30. Yacoub, S., Cukic, B., & Ammar, H. H. (2004). A scenario-based reliability analysis approach for component-based software. *IEEE Transactions on Reliability*, 53(4), 465–480.
31. Hamlet, D., Mason, D., & Woit, D. (2001, July). Theory of software reliability based on components. In *Proceedings of the 23rd International Conference on Software Engineering*, pp. 361–370, IEEE Computer Society.
32. Kan, S. H. (2002). *Metrics and Models in Software Quality Engineering*, Addison-Wesley Longman Publishing Co., Inc.
33. Leblanc, S. P., & Roman, P. A. (2002). Reliability estimation of hierarchical software systems. In *Proceedings of Annual Reliability and Maintainability Symposium, 2002*, pp. 249–253, IEEE.

34. Musa, J. D., & Okumoto, K. (1984, March). A logarithmic Poisson execution time model for software reliability measurement. In *Proceedings of the 7th International Conference on SOFTWARE Engineering*, Orlando, FL, pp. 230–238, IEEE Press, Piscataway, NJ.
35. Pham, H. (2000) *Software Reliability*, Springer-Verlag, London.
36. Jelinski, Z., & Moranda, P. (1972). Software reliability research. In W. Freiberger, Ed., *Statistical Computer Performance Evaluation*, Academic Press, New York, pp. 465–484.
37. Littlewood, B., & Verrall, J. L. (1973). A Bayesian reliability growth model for computer software. *Applied Statistics*, 22(3), 332–346.
38. Goel, A. L., & Okumoto, K. (1979). A time-dependent error-detection rate model for software reliability and other performance measure. *IEEE Transactions on Reliability*, 28(3), 206–211.

12
Reliability and Fault Tolerance Modeling of Multiphase Traction Electric Motors

Ilia Frenkel, Lev Khvatskin, and Ehud Ikar
Shamoon College of Engineering

Igor Bolvashenkov and Hans-Georg Herzog
Technical University of Munich

Anatoly Lisnianski
The Israel Electric Corporation

CONTENTS
12.1 Introduction 267
12.2 Brief Description of the L_z-Transform Method 269
12.3 Multistate Model of the MPSTD 270
    12.3.1 System Description 270
    12.3.2 Element Description 272
    12.3.3 Multistate Model for MPSTD 278
        12.3.3.1 Diesel-Generator's Subsystem 279
        12.3.3.2 L_z-transform Subsystem U Calculation 280
        12.3.3.3 Multistate Model for MPSTD with Three-Phase Motor...280
        12.3.3.4 Multistate Model for MPSTD with Six-Phase Motor.....282
        12.3.3.5 Multistate Model for MPSTD with Nine-Phase Motor 283
        12.3.3.6 Multistate Model for MPSTD with 12-Phase Motor...285
    12.3.4 Calculation Reliability Indices of MPSTD..................................287
12.4 Conclusion 291
References..294

12.1 Introduction
Nowadays, the complexity of modern engineering systems is increasing. The main function of such systems is to achieve the required level of

sustainable and safety operations with maximum efficient and stable fulfillment. The implementation of the specified requirements is closely related to the assessment of sustainable operation indicators of the system.

Ships' traction systems are the safety-critical systems and their operational sustainability is obligatory. Taking into consideration the ships' specific operational conditions, the following features of their electric propulsion systems are important: high autonomy of the ships' operations, availability of structural and functional redundancies, high maintainability of an electric propulsion system, possibility of repair during operation, etc.

As shown in [2], using a multiphase electric motor allows the increase of fault tolerance of safety-critical systems and takes into account the high requirements imposed to its propulsion system. In the present study, the propulsion system of an icebreaking ship, called a multipower source traction drive (MPSTD), with four diesel generators was analyzed [1]. Such a topological scheme of traction drives is widely used for transporting icebreaking ships and icebreakers.

Except edition and increase the fault tolerance of traction drive, the structural and functional redundancies of components are used. In most cases, given the traction drive's limitations of weight and dimensions, for example, in aircraft, structural redundancy is not possible; therefore, a functional redundancy option is used.

A promising variant of the practical implementation of the functional redundancy of one of the main components of the traction electric drive is a multiphase electric motor, which is discussed in detail in this chapter. Figure 12.1 schematically shows the stator topology of a

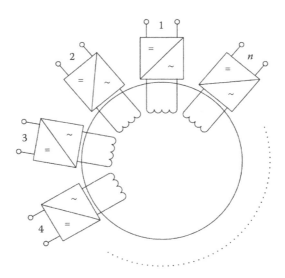

FIGURE 12.1
Stator topology of multiphase traction electric motor.

fault-tolerant multiphase permanent magnet synchronous motor. The motor phases are not galvanically connected to each other. Each phase is fed from a power source through a multilevel electrical inverter. In this chapter, the electric motors with 3-, 6-, 9-, and 12-phase topologies are considered.

Many technical systems, such as MPSTDs, are designed to perform their tasks with different performance levels: level of perfect functioning, level with reduced capacity, and complete failure level. Such systems can be described as multistate systems (MSSs). Usually, MSSs are composed of elements that can be multistate themselves. The basic concepts and the recent development of MSS reliability theory can be found in [8] and [9]. Different approaches have been introduced for the reliability analysis of such systems: direct partial logic derivatives [7] and UGF technique [11] for steady-state performance distributions and L_z-transform techniques [6] for dynamic MSS reliability analysis. Many technical applications of L_z-transform are presented in [3–6] and [11,12].

In the present chapter, the L_z-transform approach is applied to a real multistate MPSTD and its availability and performance are analyzed. It was shown that in comparison with the straightforward Markov method, the L_z-transform approach drastically simplifies the computation of the operational sustainability value for such a system.

12.2 Brief Description of the L_z-Transform Method

We consider a MSS, consisting of n multistate components. Any j-component can have k_j different states, corresponding to different performances g_{ji}, represented by the set $\mathbf{g}_j = \{g_{j1}, \ldots, g_{jk_j}\}$, $j = \{1, \ldots, n\}$; $i = \{1, 2, \ldots, k_j\}$. The performance stochastic processes $G_j(t) \in \mathbf{g}_j$ and the system structure function $G(t) = f(G_1(t), \ldots, G_n(t))$ that produce the stochastic process corresponding to the output performance of the entire MSS fully define the MSS model.

The MSS model definitions can be divided into the following steps. For each multistate component, we will build a model of the stochastic process. The Markov performance stochastic process for each component j can be represented by the expression $G_j(t) = \{\mathbf{g}_j, \mathbf{A}_j, \mathbf{p}_{j0}\}$, where $\mathbf{g}_j$ is the set of possible component's states, defined as follows: $\mathbf{A}_j = \left(a_{lm}^{(j)}(t)\right)$, $l, m = 1, \ldots, k; j = 1, \ldots, n$—transition intensities matrix and $\mathbf{p}_{j0} = \left[p_{10}^{(j)} = \Pr\{G_j(0) = g_{10}\}, \ldots, p_{k_j 0}^{(j)} = \Pr\{G_j(0) = g_{k_j 0}\}\right]$—initial states probability distribution.

For each component j, the system of Kolmogorov forward differential equations [6] can be written for the determination of the state probabilities $p_{ji}(t) = \Pr\{G_j(t) = g_{ji}\}$, $i = 1, \ldots, k_j$, $j = 1, \ldots, n$ under initial conditions $\mathbf{p}_{j0}$. Now the L_z-transform of a discrete-state continuous-time (DSCT) Markov process $G_j(t)$ for each component j can be written as follows:

$$L_Z\{G_j(t)\} = \sum_{i=1}^{k_j} p_{ji}(t) z^{g_{ji}}. \qquad (12.1)$$

In the next step, in order to find L_z-transform of the entire MSS's output performance Markov Process $G(t)$, the Ushakov's Universal Generating Operator [12] can be applied to all individual L_z-transforms $L_Z\{G_j(t)\}$ over all time points $t \geq 0$

$$L_Z\{G(t)\} = \Omega_f\{L_Z[G_1(t)],\ldots,L_Z[G_n(t)]\} = \sum_{i=1}^{K} p_i(t) z^{g_i}. \qquad (12.2)$$

The technique of Ushakov's operator application is well established for many different structure functions [7].

Using the resulting L_z-transform, MSS's mean instantaneous availability for the constant demand level w can be derived as the sum of all probabilities in the L_z-transform from terms where powers of z are not negative:

$$A(t) = \sum_{g_i \geq w} p_i(t). \qquad (12.3)$$

MSS's mean instantaneous performance may be calculated as the sum of all probabilities multiplied to performance in the L_z-transform from terms where powers of z are positive:

$$E(t) = \sum_{g_i > 0} p_i(t) g_i. \qquad (12.4)$$

The instantaneous performance deficiency $D(t)$ at any time t for the constant demand w can be calculated as follows:

$$D(t) = \sum_{g_i > 0} p_i(t) \cdot \max(w - g_i, 0). \qquad (12.5)$$

12.3 Multistate Model of the MPSTD

12.3.1 System Description

We analyze a conventional diesel-electric power drive, using it in Amguema-type arctic cargo ships, based on a direct electric propulsion system. The structure of ship's diesel-electric traction drive is shown in Figure 12.2. The

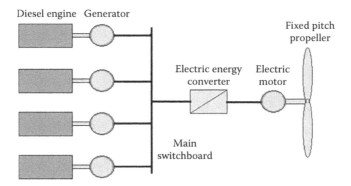

FIGURE 12.2
Structure of the ship's diesel-electric traction drive.

system consists of a diesel-generator subsystem, a main switchboard, an electric energy converter, and an electric motor.

The energy performance of the whole system is 5500 kW. Depending on the ice conditions, the amount of cargo, and other conditions of navigation, ship's propulsion system is operating with a different number of diesel and electric propulsion motors. It realizes the required value of the performance and, as a consequence, the high survivability of the ship with the possible occurrence of critical failures of power equipment.

The power-generating performance of each diesel generator is 1375 kW. Therefore, connecting a diesel generator in parallel supports the nominal generating performance, which is required for the functioning of the whole system.

The main switchboard device, the electric energy converter, and the electric motor have the nominal performance.

In the ship's diesel-electric power drives with a fix pitch propeller, the dimensions of the electric machines have to be calculated accurately in order to estimate the available sufficient propulsion power, which is directly determined by the required value of operational power and needed additional power in case of heavy weather or ice conditions in the area of navigation. Possible structures of the arctic ship's propulsion system with a different number of diesel generators and main traction motors are determined by operating conditions of the arctic ship and the ice and temperature conditions.

The typical operational modes of arctic cargo ships are as follows:

- Navigation with icebreaker in heavy ice and navigation without icebreaker in solid ice need 100% of the generated power.
- Navigation in the open water depended on the required velocity needs 75% of the generated power.

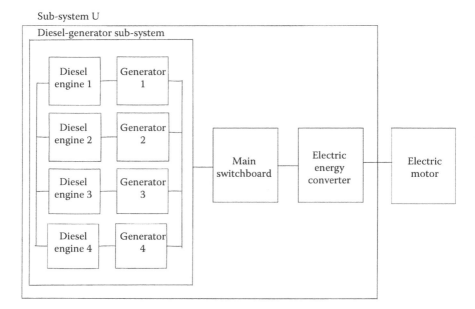

FIGURE 12.3
Reliability block diagram of the ship's diesel-electric traction drive.

As an alternative to the conventional 3-phase traction electric motor, a multiphase traction motor with 6-, 9-, and 12-phases is considered, the detailed description and features of which are presented in [2]. The reliability block diagram of the whole ship's traction drive is shown in Figure 12.3.

12.3.2 Element Description

For system's elements, which have two states (fully working and fully failed), in order to calculate the probabilities of each state we build the state-space diagram (Figure 12.4) and the following system of differential equations:

$$\begin{cases} \dfrac{dp_{i1}(t)}{dt} = -\lambda_i p_{i1}(t) + \mu_i p_{i2}(t), \\ \dfrac{dp_{i2}(t)}{dt} = \lambda_i p_{i1}(t) - \mu_i p_{i2}(t) \end{cases}$$

where i = DE, G, MS, EEC.

Initial conditions are $p_{i1}(0) = 1$; $p_{i2}(0) = 0$.

We used MATLAB® for the numerical solution of these systems of differential equations to obtain the probabilities $p_{i1}(t), p_{i2}(t)$ (i = DE, G, MS, EEC).

Multiphase Traction Electric Motors 273

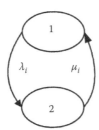

FIGURE 12.4
State-space diagram of elements with two states.

Therefore, for such system's element the output performance stochastic processes can be obtained as follows:

$$\begin{cases} \mathbf{g}_i = \{g_{i1}, g_{i1}\} = \{1375, 0\}, \\ \mathbf{p}_i(t) = \{p_{i1}(t), p_{i1}(t)\}. \end{cases}$$

Sets $\mathbf{g}_i, \mathbf{p}_i(t)$ (i = DE, G, MS, EEC) define L_z-transforms for each element as follows:

Diesel engine:

$$L_z\{g^{DE}(t)\} = p_1^{DE}(t)z^{g_1^{DE}} + p_2^{DE}(t)z^{g_2^{DE}} = p_1^{DE}(t)z^{1375} + p_2^{DE}(t)z^0. \quad (12.6)$$

Generator:

$$L_z\{g^G(t)\} = p_1^G(t)z^{g_1^G} + p_2^G(t)z^{g_2^G} = p_1^G(t)z^{1375} + p_2^G(t)z^0. \quad (12.7)$$

Main switchboard:

$$L_z\{g^{MS}(t)\} = p_1^{MS}(t)z^{g_1^{MS}} + p_2^{MS}(t)z^{g_2^{MS}} = p_1^{MS}(t)z^{5500} + p_2^{MS}(t)z^0. \quad (12.8)$$

Electric energy converter:

$$L_z\{g^{EEC}(t)\} = p_1^{EEC}(t)z^{g_{i1}^{EEC}} + p_2^{EEC}(t)z^{g_{i2}^{EEC}} = p_1^{EEC}(t)z^{5500} + p_2^{EEC}(t)z^0. \quad (12.9)$$

The system's element, three-phase motor, has three states: fully working state with performance of 5500 kW, partial failure state with performance of 3667 kW, and full failure. The state-space diagram is presented in Figure 12.5.

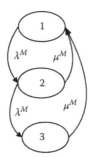

FIGURE 12.5
State-space diagram of three-phase motor.

To calculate the probabilities of each state, we build the following system of differential equations:

$$\begin{cases} \dfrac{dp_1^{M_3}(t)}{dt} = -\lambda_M p_1^{M_3}(t) + \mu_M \left(p_2^{M_3}(t) + p_3^{M_3}(t) \right), \\[6pt] \dfrac{dp_2^{M_3}(t)}{dt} = \lambda_M p_1^{M_3}(t) - (\lambda_M + \mu_M) p_2^{M_3}(t) \\[6pt] \dfrac{dp_3^{M_3}(t)}{dt} = \lambda_M p_2^{M_3}(t) - \mu_M p_3^{M_3}(t). \end{cases}$$

The initial conditions are as follows:

$$p_1^{M_3}(0) = 1;\ p_2^{M_3}(0) = 0;\ p_3^{M_3}(0) = 0.$$

We used MATLAB® for numerical solution of this system of differential equations to obtain the probabilities $p_1^{M_3}(t)$, $p_2^{M_3}(t)$, $p_3^{M_3}(t)$. Therefore, for such system's element the output performance stochastic processes can be obtained as follows:

$$\begin{cases} \mathbf{g}^{M_3} = \left\{ g_1^{M_3}, g_2^{M_3}, g_3^{M_3} \right\} = \{5500, 3670, 0\}, \\[4pt] \mathbf{p}^{M_3}(t) = \left\{ p_1^{M_3}(t), p_2^{M_3}(t), p_3^{M_3}(t) \right\}. \end{cases}$$

Sets $\mathbf{g}^{M_3}, \mathbf{p}^{M_3}(t)$ define L_z-transforms for three-phase Motor as follows:

$$L_z\left\{ \mathbf{g}^{M_3}(t) \right\} = p_1^{M_3}(t) z^{g_1^{M_3}} + p_2^{M_3}(t) z^{g_2^{M_3}} + p_3^{M_3}(t) z^{g_3^{M_3}}$$

$$= p_1^{M_3}(t) z^{5500} + p_2^{M_3}(t) z^{3670} + p_3^{M_3}(t) z^0. \quad (12.10)$$

The system's element, six-phase motor, has four states: fully working state with performance 5500 kW, partial failure states with performances 4583 and

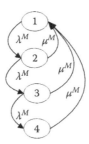

FIGURE 12.6
State-space diagram of six-phase motor.

3670 kW, and full failure. The state-space diagram is presented in Figure 12.6. To calculate the probabilities of each state we build the following system of differential equations:

$$\begin{cases} \dfrac{dp_1^{M_6}(t)}{dt} = -\lambda_M p_1^{M_6}(t) + \mu_M \left(p_2^{M_6}(t) + p_3^{M_6}(t) + p_4^{M_6}(t) \right), \\[6pt] \dfrac{dp_2^{M_6}(t)}{dt} = \lambda_M p_1^{M_6}(t) - (\lambda_M + \mu_M) p_2^{M_6}(t) \\[6pt] \dfrac{dp_3^{M_6}(t)}{dt} = \lambda_M p_2^{M_6}(t) - (\lambda_M + \mu_M) p_3^{M_6}(t) \\[6pt] \dfrac{dp_4^{M_6}(t)}{dt} = \lambda_M p_3^{M_6}(t) - \mu_M p_4^{M_6}(t). \end{cases}$$

The initial conditions are as follows:

$$p_1^{M_6}(0) = 1;\ p_2^{M_6}(0) = 0;\ p_3^{M_6}(0) = 0;\ p_4^{M_6}(0) = 0.$$

We used MATLAB® for numerical solution of this system of differential equations to obtain the probabilities $p_1^{M_6}(t),\ p_2^{M_6}(t),\ p_3^{M_6}(t),\ p_4^{M_6}(t)$. Therefore, for such system's element the output performance stochastic processes can be obtained as follows:

$$\begin{cases} \mathbf{g}^{M_6} = \left\{ g_1^{M_6}, g_2^{M_6}, g_3^{M_6}, g_4^{M_6} \right\} = \{5500, 4583, 3670, 0\}, \\ \mathbf{p}^{M_6}(t) = \left\{ p_1^{M_6}(t), p_2^{M_6}(t), p_3^{M_6}(t), p_4^{M_6}(t) \right\}. \end{cases}$$

Sets $\mathbf{g}^{M_6}, \mathbf{p}^{M_6}(t)$ define L_z-transforms for six-phase Motor as follows:

$$\begin{aligned} L_z\{\mathbf{g}^{M_6}(t)\} &= p_1^{M_6}(t) z^{g_1^{M_6}} + p_2^{M_6}(t) z^{g_2^{M_6}} + p_3^{M_6}(t) z^{g_3^{M_6}} + p_4^{M_6}(t) z^{g_4^{M_6}} \\ &= p_1^{M_6}(t) z^{5500} + p_2^{M_6}(t) z^{4583} + p_3^{M_6}(t) z^{3670} + p_4^{M_6}(t) z^0. \end{aligned} \quad (12.11)$$

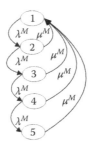

FIGURE 12.7
State-space diagram of nine-phase motor.

The system's element, nine-phase motor, has five states: fully working state with performance 5500 kW, partial failure states with performances 4889, 4278, and 3670 KW and full failure. The state-space diagram is presented in Figure 12.7. To calculate the probabilities of each state we build the following system of differential equations:

$$\begin{cases} \dfrac{dp_1^{M_9}(t)}{dt} = -\lambda_M p_1^{M_9}(t) + \mu_M \left(p_2^{M_9}(t) + p_3^{M_9}(t) + p_4^{M_9}(t) + p_5^{M_9}(t) \right), \\[4pt] \dfrac{dp_2^{M_9}(t)}{dt} = \lambda_M p_1^{M_9}(t) - (\lambda_M + \mu_M) p_2^{M_9}(t) \\[4pt] \dfrac{dp_3^{M_9}(t)}{dt} = \lambda_M p_2^{M_9}(t) - (\lambda_M + \mu_M) p_3^{M_9}(t) \\[4pt] \dfrac{dp_4^{M_9}(t)}{dt} = \lambda_M p_3^{M_9}(t) - (\lambda_M + \mu_M) p_4^{M_9}(t) \\[4pt] \dfrac{dp_5^{M_9}(t)}{dt} = \lambda_M p_4^{M_9}(t) - \mu_M p_5^{M_9}(t). \end{cases}$$

The initial conditions are as follows:

$$p_1^{M_9}(0) = 1;\ p_2^{M_9}(0) = 0;\ p_3^{M_9}(0) = 0;\ p_4^{M_9}(0) = 0;\ p_5^{M_9}(0) = 0.$$

We used MATLAB® for numerical solution of this system of differential equations to obtain probabilities $p_1^{M_9}(t)$, $p_2^{M_9}(t)$, $p_3^{M_9}(t)$, $p_4^{M_9}(t)$, $p_5^{M_9}(t)$. Therefore, for such system's element the output performance stochastic processes can be obtained as follows:

$$\begin{cases} \mathbf{g}^{M_9} = \left\{ g_1^{M_9}, g_2^{M_9}, g_3^{M_9}, g_4^{M_9}, g_5^{M_9} \right\} = \{5500, 4889, 4278, 3670, 0\}, \\[4pt] \mathbf{p}^{M_9}(t) = \left\{ p_1^{M_9}(t), p_2^{M_9}(t), p_3^{M_9}(t), p_4^{M_9}(t), p_5^{M_9}(t) \right\}. \end{cases}$$

Sets $\mathbf{g}^{M_9}, \mathbf{p}^{M_9}(t)$ define L_z-transforms for a nine-phase motor as follows:

$$L_z\{g^{M_9}(t)\} = p_1^{M_9}(t)z^{g_1^{M_9}} + p_2^{M_9}(t)z^{g_2^{M_9}} + p_3^{M_9}(t)z^{g_3^{M_9}} + p_4^{M_9}(t)z^{g_4^{M_9}} + p_5^{M_9}(t)z^{g_5^{M_9}}$$

$$= p_1^{M_9}(t)z^{5500} + p_2^{M_9}(t)z^{4889} + p_3^{M_9}(t)z^{4278} + p_4^{M_9}(t)z^{3670} + p_5^{M_9}(t)z^{0}. \quad (12.12)$$

The system's element, 12-phase motor, has six states: fully working state with performance 5500 kW, partial failure states with performances 5042, 4583, 4125, and 3670 kW, and full failure. The state-space diagram is presented in Figure 12.8. To calculate the probabilities of each state we build the following system of differential equations:

$$\begin{cases} \dfrac{dp_1^{M_{12}}(t)}{dt} = -\lambda_M p_1^{M_{12}}(t) + \mu_M \left(p_2^{M_{12}}(t) + p_3^{M_{12}}(t) + p_4^{M_{12}}(t) + p_5^{M_{12}}(t) + p_6^{M_{12}}(t) \right), \\[4pt] \dfrac{dp_2^{M_{12}}(t)}{dt} = \lambda_M p_1^{M_{12}}(t) - (\lambda_M + \mu_M) p_2^{M_{12}}(t) \\[4pt] \dfrac{dp_3^{M_{12}}(t)}{dt} = \lambda_M p_2^{M_{12}}(t) - (\lambda_M + \mu_M) p_3^{M_{12}}(t) \\[4pt] \dfrac{dp_4^{M_{12}}(t)}{dt} = \lambda_M p_3^{M_{12}}(t) - (\lambda_M + \mu_M) p_4^{M_{12}}(t) \\[4pt] \dfrac{dp_5^{M_{12}}(t)}{dt} = \lambda_M p_4^{M_{12}}(t) - (\lambda_M + \mu_M) p_5^{M_{12}}(t) \\[4pt] \dfrac{dp_6^{M_{12}}(t)}{dt} = \lambda_M p_5^{M_{12}}(t) - \mu_M p_6^{M_{12}}(t). \end{cases}$$

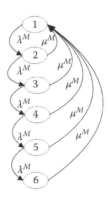

FIGURE 12.8
State-space diagram of 12-phase motor.

The initial conditions are as follows:

$$p_1^{M_{12}}(0) = 1; \; p_2^{M_{12}}(0) = p_3^{M_{12}}(0) = p_4^{M_{12}}(0) = p_5^{M_{12}}(0) = p_6^{M_{12}}(0) = 0.$$

We used MATLAB® for numerical solution of this system of differential equations to obtain the probabilities $p_1^{M_{12}}(t)$, $p_2^{M_{12}}(t)$, $p_3^{M_{12}}(t)$, $p_4^{M_{12}}(t)$, $p_5^{M_{12}}(t)$, $p_6^{M_{12}}(t)$. Therefore, for such system's element the output performance stochastic processes can be obtained as follows:

$$\begin{cases} \mathbf{g}^{M_{12}} = \left\{ g_1^{M_{12}}, g_2^{M_{12}}, g_3^{M_{12}}, g_4^{M_{12}}, g_5^{M_{12}}, g_6^{M_{12}} \right\} = \{5500, 5042, 4583, 4125, 3670, 0\}, \\ \mathbf{p}^{M_{12}}(t) = \left\{ p_1^{M_{12}}(t), p_2^{M_{12}}(t), p_3^{M_{12}}(t), p_4^{M_{12}}(t), p_5^{M_{12}}(t), p_6^{M_{12}}(t) \right\}. \end{cases}$$

Sets $\mathbf{g}^{M_{12}}$, $\mathbf{p}^{M_{12}}(t)$ define L_z-transforms for nine-phase Motor as follows:

$$L_z\left\{g^{M_{12}}(t)\right\} = p_1^{M_{12}}(t) z^{g_1^{M_{12}}} + p_2^{M_{12}}(t) z^{g_2^{M_{12}}} + p_3^{M_{12}}(t) z^{g_3^{M_{12}}} + p_4^{M_{12}}(t) z^{g_4^{M_{12}}}$$

$$+ p_5^{M_{12}}(t) z^{g_5^{M_{12}}} + p_6^{M_{12}}(t) z^{g_6^{M_{12}}}$$

$$= p_1^{M_{12}}(t) z^{5500} + p_2^{M_{12}}(t) z^{5042} + p_3^{M_{12}}(t) z^{4583} + p_4^{M_{12}}(t) z^{4125}$$

$$+ p_5^{M_{12}}(t) z^{3670} + p_6^{M_{12}}(t) z^0. \tag{12.13}$$

12.3.3 Multistate Model for MPSTD

As is shown in Figure 12.3, the multistate model for MPSTD may be presented as connected in series diesel-generator subsystem, main switchboard, electric energy converter, and electric motor. For simplification, we will calculate the whole-system L_z-transform separately: first, L_z-transform of the Diesel-Generator subsystem, second, L_z-transform of connected in series diesel-generator subsystem, main switchboard, electric energy converter (this subsystem is names subsystem U) and third, whole system with different kinds of electric motors. Therefore, the whole-system L_z-transform is as follows:

$$L_z\left\{G^{MPSTD}(t)\right\} = \Omega_{f_{ser}}\left(L_z\left\{G^{DGS}(t)\right\}, L_z\left\{G^{MS}(t)\right\}, L_z\left\{G^{EEC}(t)\right\}, L_z\left\{G^{M}(t)\right\}\right)$$

$$= \Omega_{f_{ser}}\left(\Omega_{f_{ser}}\left(L_z\left\{G^{DGS}(t)\right\}, L_z\left\{G^{MS}(t)\right\}, L_z\left\{G^{EEC}(t)\right\}\right), L_z\left\{G^{M}(t)\right\}\right)$$

$$= \Omega_{f_{ser}}\left(L_z\left\{G^{U}(t)\right\}, L_z\left\{G^{M}(t)\right\}\right). \tag{12.14}$$

12.3.3.1 Diesel-Generator's Subsystem

A diesel-generator subsystem consists of four identical pairs of diesel engines and generators connected in parallel. Each diesel engine and each generator is a two-state device: power generating performance of a fully operational state is 1375 kW and a total failure corresponds to a capacity of 0.

Using the composition operator Ω_{fser}, we obtain the L_z-transform $L_z\{G^{DG}(t)\}$ for each pair of identical diesel engines and generators, connected in series, where the powers of z are found as a minimum of powers of corresponding terms:

$$L_z\{G^{DG}(t)\} = \Omega_{fser}\left(g^{DE}(t), g^G(t)\right)$$
$$= p_1^{DE}(t) p_1^G(t) z^{1375} + \left(p_1^{DE}(t) p_2^G(t) + p_2^{DE}(t)\right) z^0. \quad (12.15)$$

Using the following notations:

$$p_1^{DG}(t) = p_1^{DE}(t) p_1^G(t)$$
$$p_2^{DG}(t) = p_1^{DE}(t) p_2^G(t) + p_2^{DE}(t),$$

we obtain the resulting L_z-transform for the diesel-generator subsystem in the following form:

$$L_z\{G^{DG}(t)\} = p_1^{DG}(t) z^{1375} + p_2^{DG}(t) z^0. \quad (12.16)$$

Using the composition operator Ω_{fpar} for 4 diesel-generators, connected in parallel, we obtain the L_z-transform $L_z\{G^{SysDG}(t)\}$ for the whole diesel-generator subsystem as follows:

$$L_z\{G^{DGS}(t)\} = \Omega_{fpar}\left(L_z\{G^{DG}(t)\}, L_z\{G^{DG}(t)\}, L_z\{G^{DG}(t)\}, L_z\{G^{DG}(t)\}\right). \quad (12.17)$$

Using notations

$$P_1^{DGS}(t) = \{p_1^{DG}(t)\}^4,$$
$$P_2^{DGS}(t) = 4 \cdot \{p_1^{DG}(t)\}^3 p_2^{DG}(t),$$
$$P_3^{DGS}(t) = 6 \cdot \{p_1^{DG}(t)\}^2 \{p_2^{DG}(t)\}^2,$$
$$P_4^{DGS}(t) = 4 \cdot p_1^{DG}(t) \{p_2^{DG}(t)\}^3,$$
$$P_5^{DGS}(t) = \{p_2^{DG}(t)\}^4,$$

we obtain the resulting L_z-transform for the whole diesel-generator subsystem in the following form:

$$L_z\{G^{DGS}(t)\} = P_1^{DGS}(t)z^{5500} + P_2^{DGS}(t)z^{4125} + P_3^{DGS}(t)z^{2750}$$
$$+ P_4^{DGS}(t)z^{1375} + P_5^{DGS}(t)z^0. \qquad (12.18)$$

12.3.3.2 L_z-transform Subsystem U Calculation

Using the composition operator Ω_{fser} for connected in series diesel-generator subsystem, main switchboard, electric energy converter, we obtain the L_z-transform $L_z\{G^U(t)\}$, where the powers of z are found as minimum of powers of corresponding terms:

$$L_z\{G^U(t)\} = \Omega_{fser}\left(L_z\{G^{DGS}(t)\}, L_z\{G^{MS}(t)\}, L_z\{G^{EEC}(t)\}\right)$$
$$= \Omega_{fser}\left(P_1^{DGS}(t)z^{5500} + P_2^{DGS}(t)z^{4125} + P_3^{DGS}(t)z^{2750} + P_4^{DGS}(t)z^{1375}\right.$$
$$\left. + P_5^{DGS}(t)z^0, p_1^{MS}(t)z^{5500} + p_2^{MS}(t)z^0, p_1^{EEC}(t)z^{5500} + p_2^{EEC}(t)z^0\right). $$
$$(12.19)$$

Using simple algebra calculations of the powers of z as minimum values of powers of corresponding terms, the whole-system L_z-transform expression is as follows:

$$L_z\{G^U(t)\} = P_1^U(t)z^{5500} + P_2^U(t)z^{4125} + P_3^U(t)z^{2750} + P_4^U(t)z^{1375} + P_5^U(t)z^0, \quad (12.20)$$

where

$$P_1^U(t) = P_1^{DGS}(t)p_1^{MS}(t)p_1^{EEC}(t),$$
$$P_2^U(t) = P_2^{DGS}(t)p_1^{MS}(t)p_1^{EEC}(t),$$
$$P_3^U(t) = P_3^{DGS}(t)p_1^{MS}(t)p_1^{EEC}(t),$$
$$P_4^U(t) = P_4^{DGS}(t)p_1^{MS}(t)p_1^{EEC}(t),$$
$$P_5^U(t) = P_5^{DGS}(t)p_1^{MS}(t)p_1^{EEC}(t) + p_2^{MS}(t)p_1^{EEC}(t) + p_2^{EEC}(t).$$

12.3.3.3 Multistate Model for MPSTD with Three-Phase Motor

The state-transition diagram of the diesel-electric power drive is presented in Figure 12.9.

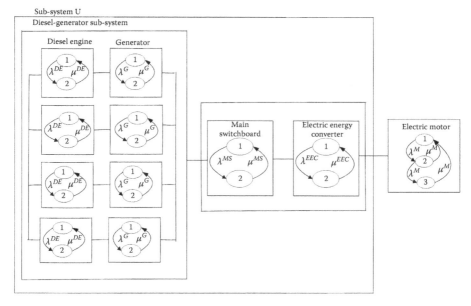

FIGURE 12.9
State-transition diagram of the MPSTD with three-phase traction electric motor.

Using the composition operator Ω_{fser} for connected in series subsystem U and three-phase motor, we obtain the L_z-transform $L_z\{G^{SysM3}(t)\}$, where the powers of z are found as minimum of powers of corresponding terms:

$$L_z\{G^{SysM3}(t)\} = \Omega_{fser}\left(L_z\{G^U(t)\}, L_z\{g^{M_3}(t)\}\right)$$

$$= \Omega_{fser}\left(P_1^U(t)z^{5500} + P_2^U(t)z^{4125} + P_3^U(t)z^{2750} + P_4^U(t)z^{1375} + P_5^U(t)z^0,\right.$$

$$\left. p_1^{M_3}(t)z^{5500} + p_2^{M_3}(t)z^{3670} + p_3^{M_3}(t)z^0\right). \tag{12.21}$$

Using simple algebra calculations of the powers of z, the whole-system L_z-transform expression is as follows:

$$L_z\{G^{SysM3}(t)\} = \sum_{i=1}^{6} P_i^{SysM3}(t)z^{g_i^{SysM3}}$$

$$= P_1^{SysM3}(t)z^{5500} + P_2^{SysM3}(t)z^{4125} + P_3^{SysM3}(t)z^{3670}$$

$$+ P_4^{SysM3}(t)z^{2750} + P_5^{SysM3}(t)z^{1375} + P_6^{SysM3}(t)z^0 \tag{12.22}$$

where

$$P_1^{SysM3}(t) = P_1^U(t) p_1^{M_3}(t),$$

$$P_2^{SysM3}(t) = P_2^U(t) p_1^{M_3}(t),$$

$$P_3^{SysM3}(t) = \left(P_1^U(t) + P_2^U(t)\right) p_2^{M_3}(t),$$

$$P_4^{SysM3}(t) = P_3^U(t)\left(p_1^{M_3}(t) + p_2^{M_3}(t)\right),$$

$$P_5^{SysM3}(t) = P_4^U(t)\left(p_1^{M_3}(t) + p_2^{M_3}(t)\right),$$

$$P_6^{SysM3}(t) = P_3^{M_3}(t) + P_5^U(t)\left(p_1^{M_3}(t) + p_2^{M_3}(t)\right).$$

12.3.3.4 Multistate Model for MPSTD with Six-Phase Motor

The state-transition diagram of the multipower drive is presented in Figure 12.10.

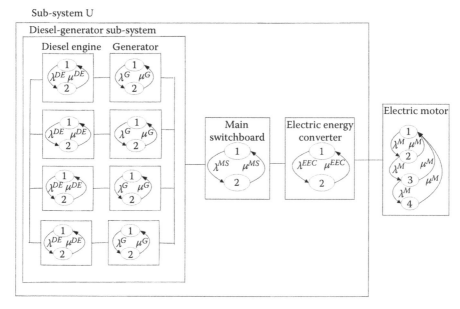

FIGURE 12.10
State-transition diagram of the MPSTD with six-phase traction electric motor.

Using the composition operator Ω_{fser} for connected in series subsystem U and six-phase motor, we obtain the L_z-transform $L_z\{G^{SysM6}(t)\}$, where the powers of z are found as the minimum of powers of corresponding terms:

$$L_z\{G^{SysM6}(t)\} = \Omega_{fser}\left(L_z\{G^U(t)\}, L_z\{g^{M6}(t)\}\right)$$

$$= \Omega_{fser}\left(P_1^U(t)z^{5500} + P_2^U(t)z^{4125} + P_3^U(t)z^{2750} + P_4^U(t)z^{1375} + P_5^U(t)z^0,\right.$$

$$\left. p_1^{M6}(t)z^{5500} + p_2^{M6}(t)z^{4583} + p_3^{M6}(t)z^{3670} + p_4^{M6}(t)z^0\right). \quad (12.23)$$

The whole-system L_z-transform expression is as follows:

$$L_z\{G^{SysM6}(t)\} = \sum_{i=1}^{7} P_i^{SysM6}(t) z^{g_i^{SysM6}}$$

$$= P_1^{SysM6}(t)z^{5500} + P_2^{SysM6}(t)z^{4583} + P_3^{SysM6}(t)z^{4125} + P_4^{SysM6}(t)z^{3670}$$

$$+ P_5^{SysM6}(t)z^{2570} + P_6^{SysM6}(t)z^{1375} + P_7^{SysM6}(t)z^0 \quad (12.24)$$

where

$$P_1^{SysM6}(t) = P_1^U(t) p_1^{M6}(t),$$

$$P_2^{SysM6}(t) = P_1^U(t) p_2^{M6}(t),$$

$$P_3^{SysM6}(t) = P_2^U(t)\left(p_1^{M6}(t) + p_2^{M6}(t)\right),$$

$$P_4^{SysM6}(t) = \left(P_1^U(t) + P_2^U(t)\right) p_3^{M6}(t),$$

$$P_5^{SysM6}(t) = P_3^U(t)\left(1 - p_4^{M6}(t)\right),$$

$$P_6^{SysM6}(t) = P_4^U(t)\left(1 - p_4^{M6}(t)\right),$$

$$P_7^{SysM6}(t) = p_4^{M6}(t) + P_5^U(t)\left(1 - p_4^{M6}(t)\right).$$

12.3.3.5 Multistate Model for MPSTD with Nine-Phase Motor

The state-transition diagram of the multipower drive is presented in Figure 12.11.

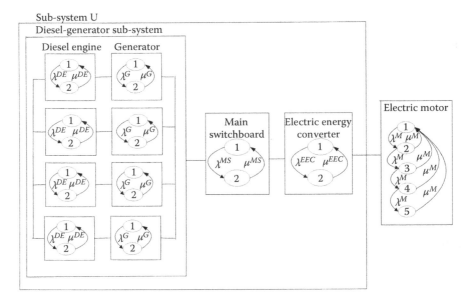

FIGURE 12.11
State-transition diagram of the MPSTD with nine-phase traction electric motor.

Using the composition operator Ω_{fser} for connected in series subsystem U and nine-phase motor, we obtain the L_z-transform $L_z\{G^{SysM9}(t)\}$, where the powers of z are found as the minimum of powers of corresponding terms:

$$L_z\{G^{SysM9}(t)\} = \Omega_{fser}\left(L_z\{G^U(t)\}, L_z\{g^{M9}(t)\}\right)$$

$$= \Omega_{fser}\left(P_1^U(t)z^{5500} + P_2^U(t)z^{4125} + P_3^U(t)z^{2750} + P_4^U(t)z^{1375} + P_5^U(t)z^0,\right.$$

$$\left. p_1^{M9}(t)z^{5500} + p_2^{M9}(t)z^{4889} + p_3^{M9}(t)z^{4278} + p_4^{M9}(t)z^{3670} + p_5^{M9}(t)z^0\right).$$
(12.25)

The whole-system L_z-transform expression is as follows:

$$L_z\{G^{SysM9}(t)\} = \sum_{i=1}^{8} P_i^{SysM9}(t)z^{g_i^{SysM9}}$$

$$= P_1^{SysM9}(t)z^{5500} + P_2^{SysM9}(t)z^{4889} + P_3^{SysM9}(t)z^{4278} + P_4^{SysM9}(t)z^{4125}$$

$$+ P_5^{SysM9}(t)z^{3670} + P_6^{SysM9}(t)z^{2750} + P_7^{SysM9}(t)z^{1375} + P_8^{SysM9}(t)z^0 \quad (12.26)$$

where

$$P_1^{SysM9}(t) = P_1^U(t) p_1^{M9}(t),$$

$$P_2^{SysM9}(t) = P_1^U(t) p_2^{M9}(t),$$

$$P_3^{SysM9}(t) = P_1^U(t) p_3^{M9}(t),$$

$$P_4^{SysM9}(t) = P_2^U(t)\left(p_1^{M9}(t) + p_2^{M9}(t) + p_3^{M9}(t)\right),$$

$$P_5^{SysM9}(t) = P_4^U(t)\left(p_1^{M9}(t) + p_2^{M9}(t)\right),$$

$$P_6^{SysM9}(t) = P_3^U(t)\left(1 - p_5^{M9}(t)\right),$$

$$P_7^{SysM9}(t) = P_4^U(t)\left(1 - p_5^{M9}(t)\right),$$

$$P_8^{SysM9}(t) = p_5^{M9}(t) + P_5^U(t)\left(1 - p_5^{M9}(t)\right).$$

12.3.3.6 Multistate Model for MPSTD with 12-Phase Motor

The state-transition diagram of the multipower drive is presented in Figure 12.12.

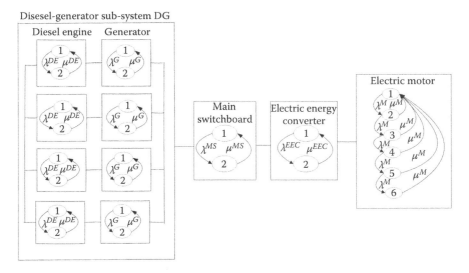

FIGURE 12.12
State-transition diagram of the MPSTD with 12 phase traction electric motor.

Using the composition operator Ω_{fser} for connected in series subsystem U and 12-phase motor, we obtain the L_z-transform $L_z\{G^{SysM12}(t)\}$, where the powers of z are found as the minimum of powers of corresponding terms:

$$L_z\{G^{SysM12}(t)\} = \Omega_{fser}\left(L_z\{G^U(t)\}, L_z\{g^{M12}(t)\}\right)$$

$$= \Omega_{fser}\left(P_1^U(t)z^{5500} + P_2^U(t)z^{4125} + P_3^U(t)z^{2750} + P_4^U(t)z^{1375} + P_5^U(t)z^0,\right.$$

$$p_1^{M12}(t)z^{5500} + p_2^{M12}(t)z^{5042} + p_3^{M12}(t)z^{4583} + p_4^{M12}(t)z^{4125}$$

$$\left. + p_5^{M12}(t)z^{3670} + p_6^{M12}(t)z^0\right). \tag{12.27}$$

The whole-system L_z-transform expression is as follows:

$$L_z\{G^{SysM12}(t)\} = \sum_{i=1}^{8} P_i^{SysM12}(t)z^{g_i^{SysM12}}$$

$$= P_1^{SysM12}(t)z^{5500} + P_2^{SysM12}(t)z^{5042} + P_3^{SysM12}(t)z^{4583} + P_4^{SysM12}(t)z^{4125}$$

$$+ P_5^{SysM12}(t)z^{3670} + P_6^{SysM12}(t)z^{2750} + P_7^{SysM12}(t)z^{1375} + P_8^{SysM12}(t)z^0$$
(12.28)

where

$$P_1^{SysM12}(t) = P_1^U(t)p_1^{M12}(t),$$

$$P_2^{SysM12}(t) = P_1^U(t)p_2^{M12}(t),$$

$$P_3^{SysM12}(t) = P_1^U(t)p_3^{M12}(t),$$

$$P_4^{SysM12}(t) = P_2^U(t)\left(p_1^{M12}(t) + p_2^{M12}(t) + p_3^{M12}(t) + p_4^{M12}(t)\right) + P_1^U(t)p_4^{M12}(t),$$

$$P_5^{SysM12}(t) = \left(P_1^U(t) + P_2^U(t)\right)p_5^{M12}(t),$$

$$P_6^{SysM12}(t) = P_3^U(t)\left(1 - p_6^{M12}(t)\right),$$

$$P_7^{SysM12}(t) = P_4^U(t)\left(1 - p_6^{M12}(t)\right),$$

$$P_8^{SysM12}(t) = p_5^U(t) + P_6^{M12}(t)\left(1 - p_5^U(t)\right).$$

12.3.4 Calculation Reliability Indices of MPSTD

Using expression (12.3), the MSS instantaneous availability of the MPSTD for different constant demand levels w may be presented as follows:

- For 100% demand level ($w = 5500\,\text{kW}$)

$$A^{SysM3}_{w \geq 5500\,\text{kW}}(t) = \sum_{g_i^{SysM3} \geq 5500} P_i^{SysM3}(t) = P_1^{SysM3}(t),$$

$$A^{SysM6}_{w \geq 5500\,\text{kW}}(t) = \sum_{g_i^{SysM6} \geq 5500} P_i^{SysM6}(t) = P_1^{SysM6}(t),$$

$$A^{SysM9}_{w \geq 5500\,\text{kW}}(t) = \sum_{g_i^{SysM9} \geq 5500} P_i^{SysM9}(t) = P_1^{SysM9}(t), \qquad (12.29)$$

$$A^{SysM12}_{w \geq 5500\,\text{kW}}(t) = \sum_{g_i^{SysM12} \geq 5500} P_i^{SysM12}(t) = P_1^{SysM12}(t).$$

- For 75% demand level ($w = 4125\,\text{kW}$)

$$A^{SysM3}_{w \geq 4125\,\text{kW}}(t) = \sum_{g_i^{SysM3} \geq 4125} P_i^{SysM3}(t) = \sum_{i=1}^{2} P_i^{SysM3}(t),$$

$$A^{SysM6}_{w \geq 4125\,\text{kW}}(t) = \sum_{g_i^{SysM6} \geq 4125} P_i^{SysM6}(t) = \sum_{i=1}^{3} P_i^{SysM6}(t),$$

$$A^{SysM9}_{w \geq 4125\,\text{kW}}(t) = \sum_{g_i^{SysM9} \geq 4125} P_i^{SysM9}(t) = \sum_{i=1}^{4} P_i^{SysM9}(t), \qquad (12.30)$$

$$A^{SysM12}_{w \geq 4125\,\text{kW}}(t) = \sum_{g_i^{SysM12} \geq 4125} P_i^{SysM12}(t) = \sum_{i=1}^{4} P_i^{SysM12}(t).$$

Using expression (12.4), the MSS instantaneous power performance of the MPSTD can be obtained as follows:

$$E^{SysM3}(t) = \sum_{g_i^{SysM3} > 0} g_i^{SysM3} P_i^{SysM3}(t) = \sum_{i=1}^{4} g_i^{SysM3} P_i^{SysM3}(t),$$

$$E^{SysM6}(t) = \sum_{g_i^{SysM6} > 0} g_i^{SysM6} P_i^{SysM6}(t) = \sum_{i=1}^{6} g_i^{SysM6} P_i^{SysM6}(t),$$

$$E^{SysM_9}(t) = \sum_{g_i^{SysM_9}>0} g_i^{SysM_9} P_i^{SysM_9}(t) = \sum_{i=1}^{7} g_i^{SysM_9} P_i^{SysM_9}(t),$$

$$E^{SysM_{12}}(t) = \sum_{g_i^{SysM_9}>0} g_i^{SysM_{12}} P_i^{SysM_{12}}(t) = \sum_{i=1}^{7} g_i^{SysM_{12}} P_i^{SysM_{12}}(t). \quad (12.31)$$

Using expression (12.5), the instantaneous power deficiency of the MPSTD for different constant demand levels can be presented as follows:

$$D_{w \geq 5500\text{kW}}^{SysM_3}(t) = \sum_{i=1}^{6} P_i^{SysM_3}(t) \cdot \max(5500 - g_i, 0)$$

$$= 1375 \cdot P_2^{SysM_3}(t) + 1830 \cdot P_3^{SysM_3}(t) + 2750 \cdot P_4^{SysM_3}(t)$$

$$+ 4125 \cdot P_5^{SysM_3}(t) + 5500 \cdot P_6^{SysM_3}(t),$$

$$D_{w \geq 5500\text{kW}}^{SysM_6}(t) = \sum_{i=1}^{7} P_i^{SysM_6}(t) \cdot \max(5500 - g_i, 0)$$

$$= 917 \cdot P_2^{SysM_6}(t) + 1375 \cdot P_3^{SysM_6}(t) + 1830 \cdot P_4^{SysM_6}(t)$$

$$+ 2750 \cdot P_5^{SysM_6}(t) + 4125 \cdot P_6^{SysM_6}(t) + 5500 \cdot P_7^{SysM_6}(t),$$

$$D_{w \geq 5500\text{kW}}^{SysM_9}(t) = \sum_{i=1}^{8} P_i^{SysM_9}(t) \cdot \max(5500 - g_i, 0)$$

$$= 611 \cdot P_2^{SysM_9}(t) + 1222 \cdot P_3^{SysM_9}(t) + 1375 \cdot P_4^{SysM_9}(t)$$

$$+ 1830 \cdot P_5^{SysM_9}(t) + 2750 \cdot P_6^{SysM_9}(t) + 4125 \cdot P_7^{SysM_9}(t) + 5500 \cdot P_8^{SysM_9}(t),$$

$$D_{w \geq 5500\text{kW}}^{SysM_{12}}(t) = \sum_{i=1}^{8} P_i^{SysM_{12}}(t) \cdot \max(5500 - g_i, 0)$$

$$= 458 \cdot P_2^{SysM_{12}}(t) + 917 \cdot P_3^{SysM_{12}}(t) + 1375 \cdot P_4^{SysM_{12}}(t)$$

$$+ 1830 \cdot P_5^{SysM_{12}}(t) + 2750 \cdot P_6^{SysM_{12}}(t) + 4125 \cdot P_7^{SysM_{12}}(t)$$

$$+ 5500 \cdot P_8^{SysM_{12}}(t), \quad (12.32)$$

$$D_{w \geq 4125\text{kW}}^{SysM3}(t) = \sum_{i=1}^{6} P_i^{SysM3}(t) \cdot \max(4125 - g_i, 0)$$

$$= 455 \cdot P_3^{SysM3}(t) + 1375 \cdot P_4^{SysM3}(t) + 2750 \cdot P_5^{SysM3}(t) + 4125 \cdot P_6^{SysM3}(t),$$

$$D_{w \geq 4125\text{kW}}^{SysM6}(t) = \sum_{i=1}^{7} P_i^{SysM6}(t) \cdot \max(4125 - g_i, 0)$$

$$= 455 \cdot P_4^{SysM6}(t) + 1375 \cdot P_5^{SysM6}(t) + 2750 \cdot P_6^{SysM6}(t) + 4125 \cdot P_7^{SysM6}(t),$$

$$D_{w \geq 4125\text{kW}}^{SysM9}(t) = \sum_{i=1}^{8} P_i^{SysM9}(t) \cdot \max(5500 - g_i, 0)$$

$$= 455 \cdot P_5^{SysM9}(t) + 1375 \cdot P_6^{SysM9}(t) + 2750 \cdot P_7^{SysM9}(t) + 4125 \cdot P_8^{SysM9}(t),$$

$$D_{w \geq 4125\text{kW}}^{SysM12}(t) = \sum_{i=1}^{8} P_i^{SysM12}(t) \cdot \max(5500 - g_i, 0)$$

$$= 455 \cdot P_5^{SysM12}(t) + 1375 \cdot P_6^{SysM12}(t) + 2750 \cdot P_7^{SysM12}(t) + 4125 \cdot P_8^{SysM12}(t),$$
(12.33)

The failure and repair rates (in year^{-1}) of each system's elements are presented in Table 12.1 below.

MSS instantaneous availability of the MPSTD for different constant demand levels is presented in Figures 12.13–12.16.

As one can see from Figure 12.13 that the instantaneous availability for 100% demand level is the same for each kind of traction electric motors and after 36.5 days of usage it is 65.2%.

Figure 12.14 shows that the instantaneous availability for 75% demand level is different for each kind of traction electric motors. After 36.5 days of usage, the instantaneous availability for 3-phase traction electric motor is 0.9216, for 6-phase traction electric motors the instantaneous availability

TABLE 12.1

Failure and Repair Rates of Each System's Elements

	Failure Rates (year^{-1})	Repair Rates (year^{-1})
Diesel engine	4.99	48.6
Generator	0.2	175
Main switchboard	0.2	440
Electric energy converter	1.5	440
Electric motor	2.7	87.7

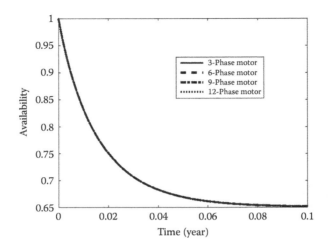

FIGURE 12.13
MSS mean instantaneous availability for 100% demand level.

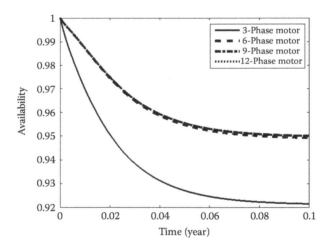

FIGURE 12.14
MSS mean instantaneous availability for 75% demand level.

is 0.9491, and for 9- and 12-phase traction electric motors the instantaneous availability is 0.95.

Comparison instantaneous availability for different demand levels for each kind of traction electric motors is presented in Figures 12.15–12.17.

Calculated MSS instantaneous power performance of the MPSTD is presented in Figure 12.18.

Calculated MSS instantaneous mean power performance deficiency of the MPSTD is presented in Figures 12.19–12.20 and in Table 12.2.

Multiphase Traction Electric Motors 291

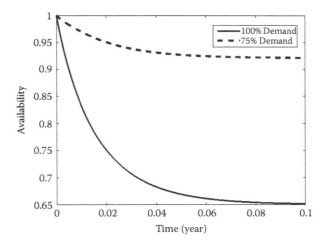

FIGURE 12.15
MSS mean instantaneous availability three-phase motor for different constant demand levels.

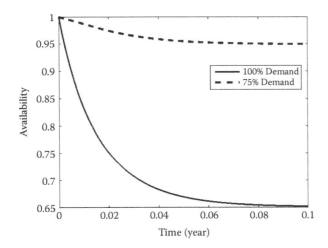

FIGURE 12.16
MSS mean instantaneous availability six-phase motors for different constant demand levels.

12.4 Conclusion

In this chapter, the L_z-transform method was used for evaluation of three important parameters of the vehicle's operational sustainability—availability, performance, and performance deficiency for different kinds of motors of the multistate MPSTD.

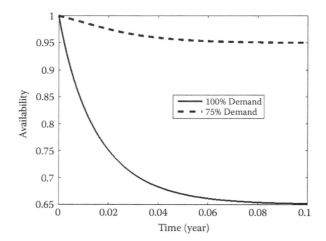

FIGURE 12.17
MSS mean instantaneous availability 9- and 12-phase motors for different constant demand levels.

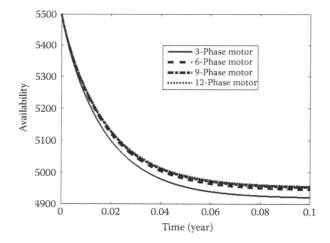

FIGURE 12.18
MSS instantaneous power performance of the MPSTD.

The L_z-transform approach extremely simplifies the solution, which in comparison with the straightforward Markov method would have required building and solving the model with 1072 states for the system with 3-phase motor, 4288 states for the system with 6-phase motor, 5375 states for the system with 9-phase motor, 6400 states for the system with 12-phase motor.

The proposed approach allows optimizing the number of the power sources of traction drive, their characteristics, and schemes of connection

Multiphase Traction Electric Motors   293

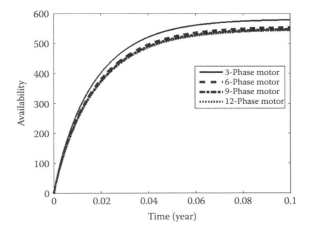

FIGURE 12.19
MSS instantaneous mean power performance deficiency for 100% demand level for different kinds of motors of the MPSTD.

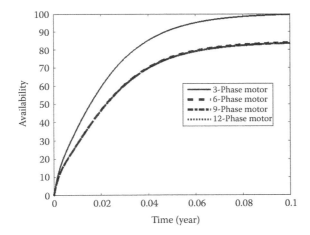

FIGURE 12.20
MSS instantaneous mean power performance deficiency for 100% demand level for different kinds of motors of the MPSTD.

TABLE 12.2

MSS Instantaneous Mean Power Performance Deficiency (in kW)

	100% Demand (kW)	75% Demand (kW)
3-Phase motor	578.3	99.8
6-Phase motor	553.8	84.1
9-Phase motor	547.6	83.7
12-Phase motor	544.1	83.7

in terms of providing the maximum operational sustainability. The results of the calculation allow concluding about the impact of the phase number of the traction electric motor on the reliability and fault tolerance indices of the entire propulsion system. Since the electrical part of the propulsion system is much more reliable than the mechanical part, an increase in the number of phases of traction electric motors does not have a significant effect on the overall performance of the propulsion system. Therefore, to improve the reliability of the entire electric drive, it is advisable to use more reliable combustion engines or their structural redundancy.

References

1. I. Bolvashenkov, H.-G. Herzog, Use of stochastic models for operational efficiency analysis of multi power source traction drives, in: *Proceedings of the Second International Symposium on Stochastic Models in Reliability Engineering, Life Science and Operations Management, (SMRLO16)*, I. Frenkel, A. Lisnianski (Eds), 15–18 February 2016, Beer Sheva, Israel, pp. 124–130.
2. I. Bolvashenkov, J. Kammermann, S. Willerich, H.-G. Herzog, Comparative study of reliability and fault tolerance of multi-phase permanent magnet synchronous motors for safety-critical drive trains, in: *Proceedings of the International Conference on Renewable Energies and Power Quality (ICREPQ16)*, 4–6 May 2016, Madrid, Spain, pp. 1–6.
3. I. Frenkel, I. Bolvashenkov, H.-G. Herzog, L. Khvatskin, Performance availability assessment of combined multi power source traction drive considering real operational conditions, *Transport and Telecommunication*, 2016, 17(3), 179–191.
4. I. Frenkel, I. Bolvashenkov, H.-G. Herzog, L. Khvatskin, Operational sustainability assessment of multi power source traction drive, in: *Mathematics Applied to Engineering*, M. Ram, J.P. Davim (Eds), Elsevier, London, 2017, pp. 191–203.
5. I. Frenkel, I. Bolvashenkov, H.-G. Herzog, L. Khvatskin, Lz-transform approach for fault tolerance assessment of various traction drives topologies of hybrid-electric helicopter, in: *Recent Advances in Multistate System Reliability: Theory and Applications*, In A. Lisnianski, I. Frenkel. A. Karagrigoriou (Eds.), Springer, London, 2017, pp. 321–362.
6. H. Jia, W. Jin, Y. Ding, Y. Song, D. Yu 2017, Multi-state time-varying reliability evaluation of smart grid with flexible demand resources utilizing Lz transform, in: *Proceedings of the International Conference on Energy Engineering and Environmental Protection (EEEP2016)*, IOP Conference Series: Earth and Environmental Science, Vol. 52, 012011, IOP Publishing, Bristol, UK.
7. M. Kvassay, E. Zaitseva, Topological Analysis of Multi-State Systems based on Direct Partial Logic Derivatives, in: *Recent Advances in Multistate System Reliability: Theory and Applications*, A. Lisnianski, I. Frenkel, & A. Karagrigoriou (Eds), Springer, London, 2018, pp. 265–281.
8. A. Lisnianski, I. Frenkel, Y. Ding, *Multi-state System Reliability Analysis and Optimization for Engineers and Industrial Managers*, Springer, London, 2010.

9. B. Natvig, *Multistate Systems Reliability, Theory with Applications*, Wiley, New York, 2011.
10. K. Trivedi, *Probability and Statistics with Reliability, Queuing and Computer Science Applications*, Wiley, New York, 2002.
11. I. Ushakov, A universal generating function, *Soviet Journal of Computer and System Sciences*, 1986, 24, 37–49.
12. H. Yu, J. Yang, H. Mo, Reliability analysis of repairable multi-state system with common bus performance sharing, *Reliability Engineering and System Safety*, 2014, 132, 90–96.

Index

A

"Absolute security," 16–17
Agglomerative hierarchical methods, 130
Altitude and orbit control system (AOCS), 68, 73, 75
   system with partially repairable components, 60–62
AMIF *see* Analytical model for interactive failures (AMIF)
Ammar model, 260
Analytical model for interactive failures (AMIF), 5
AOCS *see* Altitude and orbit control system (AOCS)

B

Barlow's model of reliability, 14–15
Bayesian model, 12
Biomass
   Gasifier system, techno-economical data of, 226
   operated generator, 218
   potential, 226
Black box, 257
Branching Poisson process (BPP), 119

C

Canonical correlation analysis, 132
CBS *see* Component-based software (CBS)
CBSD *see* Component-based software development (CBSD)
Cheung model, 259
Cluster analysis, 130–132
COA *see* Cuckoo optimization algorithm (COA)
Combinatorial methods, 48–49
Complex Poisson process (CPP), 11
Component-based software (CBS), 258
Component-based software development (CBSD), 254–262
Component-based software systems, modeling reliability of
   software failure mechanisms, 251–252
   software *versus* hardware reliability, 252–254
Condensation thermal power plants (Co-TPP)
   entries and exits toward environment, 35–37
   failures and damages during operation of, 26–31
   forecasting reliability of complex technical systems, 38–41
   growth of reliability of system, mathematical models of, 12–16
   indicators of reliability of, 18–25
   maintenance activities in, 38
   optimal reliability of, 31–34
   reliability assessment methods, 16–18
   reliability of complex technical systems, models for, 3–12
Condition-based software rejuvenation schemes, 83
Continuous-time Markov chain (CTMC), 49, 82, 258
Continuous-time stochastic process, 50
Control charts, 199, 200
Conventional diesel-electric power drive, 268–269
Correlations analysis, 122–129
Correspondence analysis procedure, 133
Corrosion-fatigue processes, 30
Cost-effectiveness analysis, 94–95
Co-TPP *see* Condensation thermal power plants (Co-TPP)
CPP *see* Complex Poisson process (CPP)
CTMC *see* Continuous-time Markov chain (CTMC)
Cuckoo optimization algorithm (COA), 191–192
Cukic model, 260

D

Diesel-generator's subsystem, 277–278
Discrete-time Markov chain (DTMC), 258
Discriminant analysis procedure, 134–137
Distribution-free monitoring scheme, 201
Doubly stochastic Poisson process (DPP), 11
DPP *see* Doubly stochastic Poisson process (DPP)
Driving forces, 109–110
DTMC *see* Discrete-time Markov chain (DTMC)
Duane's mathematical model of reliability growth, 14
Dynamic module
  with structure variation, 66–67
  without structure variation, 65–66
Dynamic non-exponential systems
  approximation approach, 50–51
  numerical example, 54–59
  simulation approach, 52–53

E

Electrical energy system (EES), 18
Empirical reliability model, 15
EOQ *see* European Quality Organization (EOQ)
ESSA *see* Extended split system approach (ESSA)
European Quality Organization (EOQ), 16
Expected total discounted cost analysis, 95–99
Exponential general-order statistic model, 167–169
Extended split system approach (ESSA), 5
Extensive simulation experiments, 172–176

F

Factor analysis, 129
Filtered Poisson process (FPP), 11
Fine reliability calculation, 17
Flame spraying, 37
FPP *see* Filtered Poisson process (FPP)

G

GA *see* Genetic algorithms (GA)
Galactic Cosmic Rays (GCR), 69
Generalized exponential general-order statistic model, 170–172
Generalized renewal processes (GRP), 119
General-order statistic (GOS) models, Bayesian inference on
  data analysis, 176–180
  extensive simulation experiments, 172–176
  model assumptions and prior selection, 166–167
  posterior analysis of different, 167–172
Genetic algorithms (GA), 188–190
German Kraftwerk Kennzeichen (KKS) system, 37
Gokhale model, 259–260
GOS models *see* General-order statistic (GOS) models
Gradual failure, 19–20
Greenhouse gas emission, 222
Grid sale/grid purchase constraint, 222
  electricity generation of, 229–230
GRP *see* Generalized renewal processes (GRP)

H

Hamlet model, 260
HESS *see* Hybrid energy storage system (HESS)
High cluster, 131
Homogeneous Poisson process (HPP), 11, 71
Hourly load demand, 223–224
Hourly solar potential, 225
HPP *see* Homogeneous Poisson process (HPP)
Hybrid energy storage system (HESS), 216
Hydroelectric power plants, 25

Index

I

Identical components
   subsystems with
      different number of, 150–151, 154–155
      equal number of, 155–157
      same number of, 151–152
IEC see International Electrotechnical Commission (IEC)
IGDT technique see Information gap decision theory (IGDT) technique
Immediate failure, 19
Information gap decision theory (IGDT) technique, 216
International Electrotechnical Commission (IEC), 109
   reliability of system with methods of, 110–120
Iterative design-develop process, 13

K

KKS system see German Kraftwerk Kennzeichen (KKS) system
k-means method, 130
Krishnamurthy model, 260

L

Laprie model, 259
Large-scale system reliability-redundancy allocation problem, 187–188
Ledoux model, 260
Littewood model, 258
Low cluster, 131
L_z-transform method, 267–268
   subsystem U calculation, 287

M

Markov chain-based approaches, 49
Markov Chain Monte Carlo (MCMC) technique, 165
Markov regenerative process (MRGP), 49
   and multistate random shocks model, 69–73
Mathur model, 260
Maximum power point tracking (MPPT), 216
MCMC technique see Markov Chain Monte Carlo (MCMC) technique
Metaheuristic algorithms, 222–223
MFT see Modularized fault tree (MFT)
Middle cluster, 131
Mini-component method, 64
Modularized fault tree (MFT), 62–63
Monte Carlo (MC) simulation method, 5, 52–53
MPPT see Maximum power point tracking (MPPT)
MPSTD see Multipower source traction drive (MPSTD)
MRGP see Markov regenerative process (MRGP)
Multipower source traction drive (MPSTD)
   calculation reliability indices of, 285–289
   multistate model
      for diesel-generator's subsystem, 277–278
      of element description, 270–276
      with nine-phase motor, 281–283
      with six-phase motor, 280–281
      of system description, 268–270
      with three-phase motor, 278–280
      with twelve-phase motor, 283–284
Multistate random shocks model, 69–73
Multivariate analysis, 121

N

New distribution-free reliability monitoring scheme
   characteristics of, 202–206
   general setup of, 201–202
NHPP see Nonhomogeneous Poisson process (NHPP)
Non-exponential distributions, 48–49
   analysis, 56–60
Nonhomogeneous Poisson process (NHPP), 11, 165

N

Nonidentical components
    subsystems with
        different number of, 149–150
        same number of, 152–153
Nonparametric predictive inference (NPI) approach, 83
NPI approach *see* Nonparametric predictive inference (NPI) approach

O

Objective function, 220–221
Operating constraints, 221–222
Optimal periodic software rejuvenation policy
    availability analysis, 92–93
    cost-effectiveness analysis, 94–95
    expected cost analysis, 87–91
    expected total discounted cost analysis, 95–99
    model description, 84–87
    numerical examples, 99–104
Optimization algorithm, 222–223

P

Partially repairable non-exponential components, 60–69
Particle swarm optimization (PSO), 190–191, 216
Path-based models, 260–261
Periodic rejuvenation, 82
PERT *see* Program Evaluation and Review Technique (PERT)
Petri-nets-based approaches, 49
Phased mission (PM) systems, 48–49
    analysis considering random shocks effect, 69–77
    with partially repairable non-exponential components, 60–69
    simulations for estimation of l, 157–160
    system operational time for exponential life times, 148–153
    system reliability function, 146–148
    unobservable subsystem operational times, 153–157

PHM *see* Proportional risk model (PHM)
Photovoltaic (PV) array, 219
    techno-economical data of, 226
PIM *see* Proportional intensity model (PIM)
Planned reliability growth model, 15–16
Planning of shutdowns of old blocks, 25
Plasma spraying, 37
PM systems *see* Phased mission (PM) systems
Poisson process, 10–11
Potential multivariate applications, for reliability analysis of repairable systems, 121–122
Power plant reliability index, 4
Power reliability constraint, 222
Predicative maintenance strategy, 3
Prevailing solution method, 191
Preventive maintenance actions, 6–7
Principal component analysis, 129
Proactive maintenance strategy, 3
Program Evaluation and Review Technique (PERT), 34
Proportional intensity model (PIM), 11
Proportional risk model (PHM), 11
PSO *see* Particle swarm optimization (PSO)
PV array *see* Photovoltaic (PV) array

R

Random shocks effect, 69–77
Rates of occurrence of failures (ROCOF), 11
Real time reliability index (RTRI), 5
Recurrent failures, 118
Regression analysis, 133–134
Reliability estimation model, 16
Repair blocks of electric traction system, 111
ROCOF *see* Rates of occurrence of failures (ROCOF)
Rough reliability calculation, 17
RTRI *see* Real time reliability index (RTRI)

Index

S

Scheuer's model of reliability, 14–15
SDLC *see* Software development life cycle (SDLC)
Semi-Markov process (SMP), 49, 69, 258
Shewhart-type control chart, 200
Ships' traction systems, 266
Shooman model, 260
Sliding window system (SWS), 233–244
SMP *see* Semi-Markov process (SMP)
Soft computing methods
    cuckoo optimization algorithm, 191–192
    genetic algorithms, 188–190
    particle swarm optimization, 190–191
Software
    failure mechanisms, 251–252
    versus hardware reliability, 252–254
Software aging, 82
Software development life cycle (SDLC), 250
Software rejuvenation, 82
Split system approach (SSA) model, 3
State-based models, 258–260
State-space-oriented model, 49, 50
Static module, 64–65
Statistical control charts, 199
Steady-state system availability, 92–93
Supplemental reliability indicators, 20
Sustainable hybrid energy system, 216–230
SWS *see* Sliding window system (SWS)
System operational time, 148–153
System reliability, 68–69
    function, 146–148
    redundancy allocation problem, 186–187

T

Techno-economical data of system components, 226–227
Three-dimensional graphical representation, 120–121
Time-based optimal software rejuvenation schedule, 83
Traditional time-based replacement policy, 12
Tree traversal algorithms, 260
Trend renewal process (TRP), 119
TRP *see* Trend renewal process (TRP)

U

UGF *see* Universal generating function (UGF)
UL *see* Unmet load (UL)
Universal generating function (UGF), 234
Unmet load (UL), 222
Unobservable subsystem operational times, 153–157
Utility grid, 219–220

W

Weibull
    general-order statistic model, 169–170
    model, 14
White box, 257
Wieren's model of reliability growth, 14

Y

Yacoub model, 260